POWER TO HUMAN

About the author

I am a power system engineer with extensive experience who focuses on applications of renewable energy and artificial intelligence to build a safe, smart, and green society. I began my work at a young age and can now proudly cite my years of expertise in the field of renewable energy solutions dating back fifteen years. In the most potent and well-known Power Systems Software, such as Matlab, PSS/E Siemens, DigSilent, PSCAD, Etap, etc., I have a stellar reputation for modeling and building electrical models of power systems and solving problems. knowledge of grid connection, grid compliance, power quality, power system planning, power system protection, and network regulation. This fantastic experience was gained from working together on several outsourced projects in European leaders of renewable energy environment. Delivering gains in overall operational efficiency and customer happiness. In the same manner I decided to share the skills in the easiest way.

Contents

1. PSCAD

1.1. General information

Main view of software

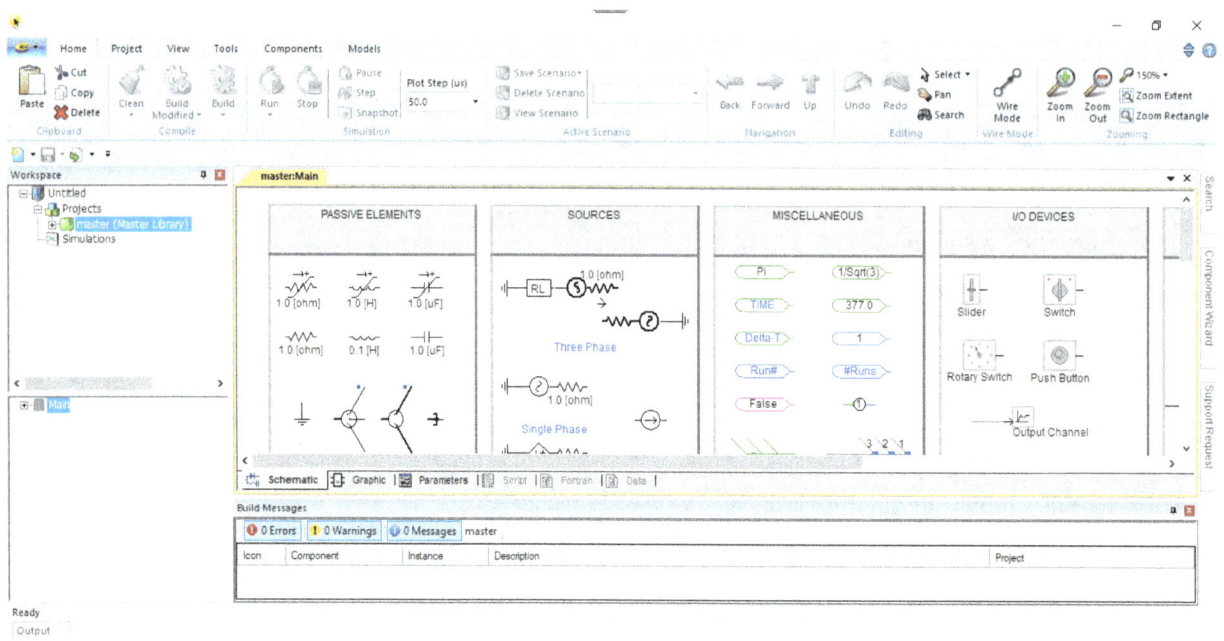

Workspace

- Untitled
  - Projects
    - master (Master Library)
  - Simulations

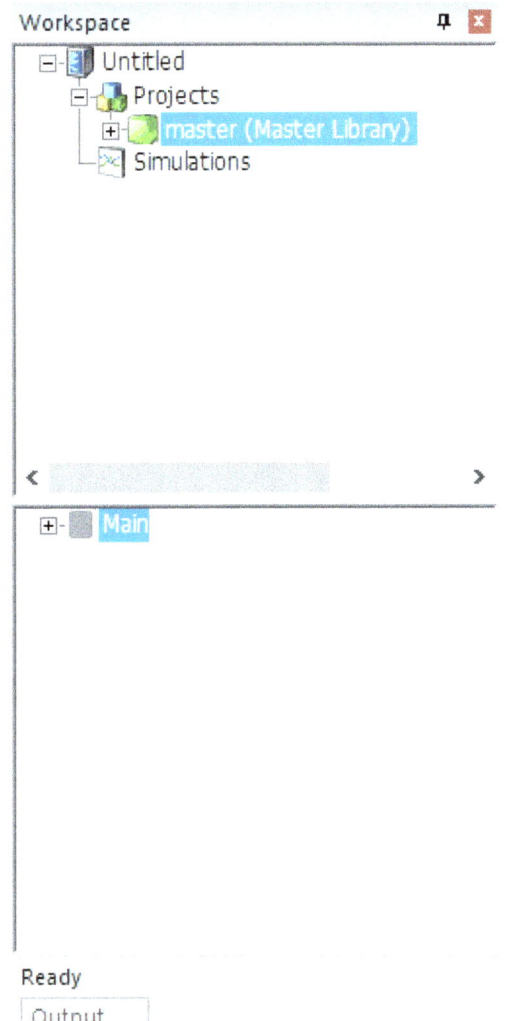

Main

Ready

Output

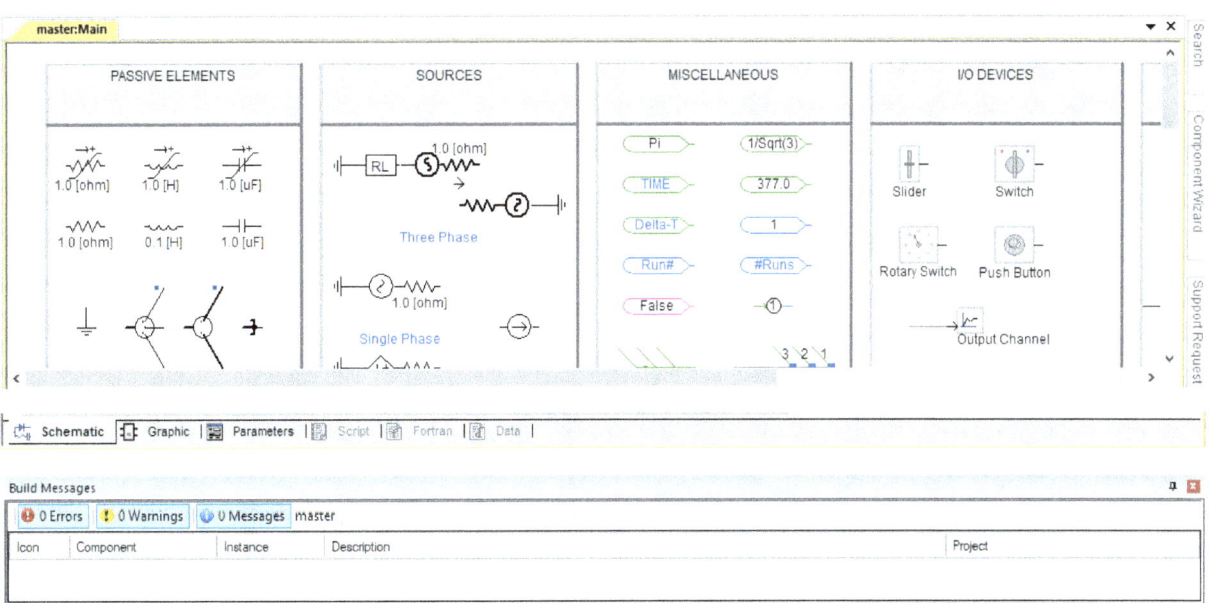

master:Main

| PASSIVE ELEMENTS | SOURCES | MISCELLANEOUS | I/O DEVICES |

PASSIVE ELEMENTS

1.0 [ohm]　　1.0 [H]　　1.0 [uF]

1.0 [ohm]　　0.1 [H]　　1.0 [uF]

SOURCES

1.0 [ohm]

RL

Three Phase

1.0 [ohm]

Single Phase

MISCELLANEOUS

Pi　　1/Sqrt(3)

TIME　　377.0

Delta-T　　1

Run#　　#Runs

False

3 2 1

I/O DEVICES

Slider　　Switch

Rotary Switch　　Push Button

Output Channel

Search

Component Wizard

Support Request

Schematic | Graphic | Parameters | Script | Fortran | Data |

Build Messages

| ❗ 0 Errors | ⚠ 0 Warnings | ❢ 0 Messages | master |

| Icon | Component | Instance | Description | Project |

1.2.   Creating new project

To create a simple project in PSCAD, click the PSCAD button in the left of the Home menu. Then head to the New and click New Project.

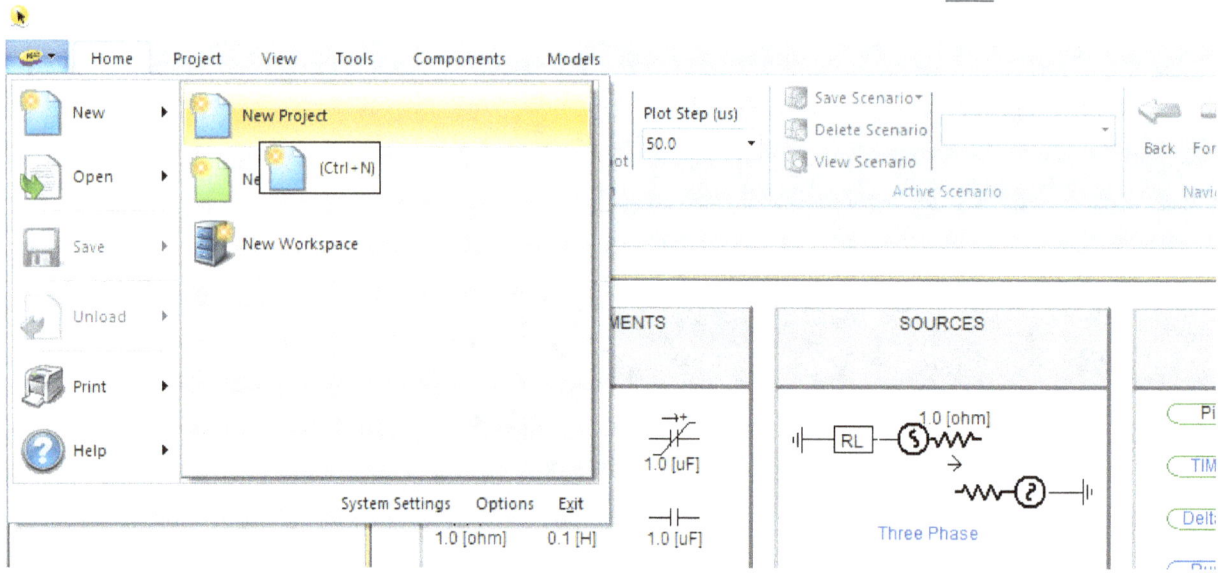

In the next following dialogue command window enter the project name and make sure that you don't insert any space. Click Ok.

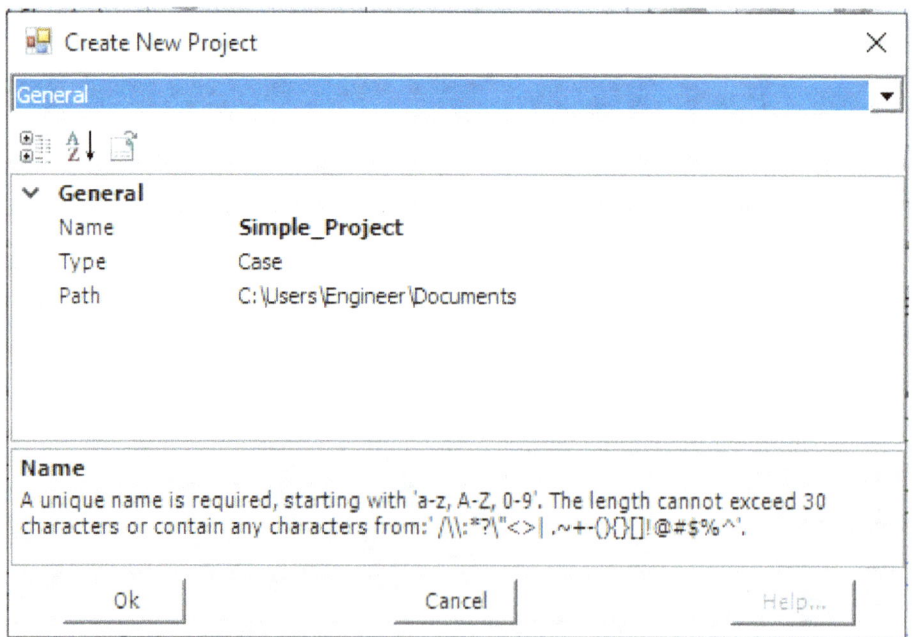

1.3.   Main elements and configurations

If we click in the middle of the blank page and zoom out with the middle mouse button we see the square working space as below.

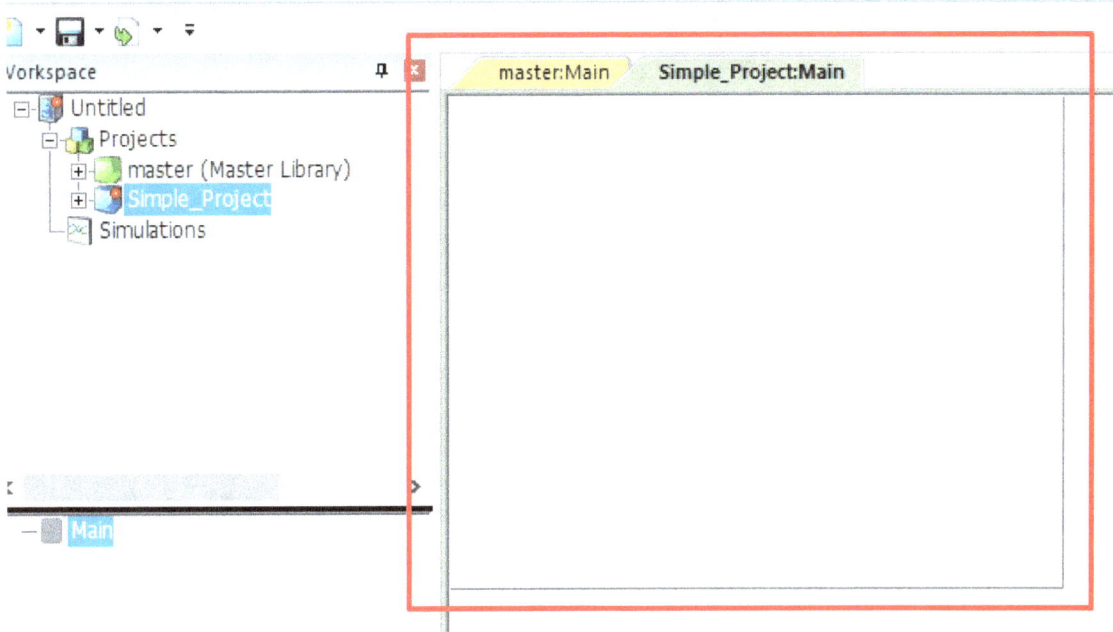

Now we will begin to analyze what settings are relevant for the user before beginning to create circuit and perform simulations.

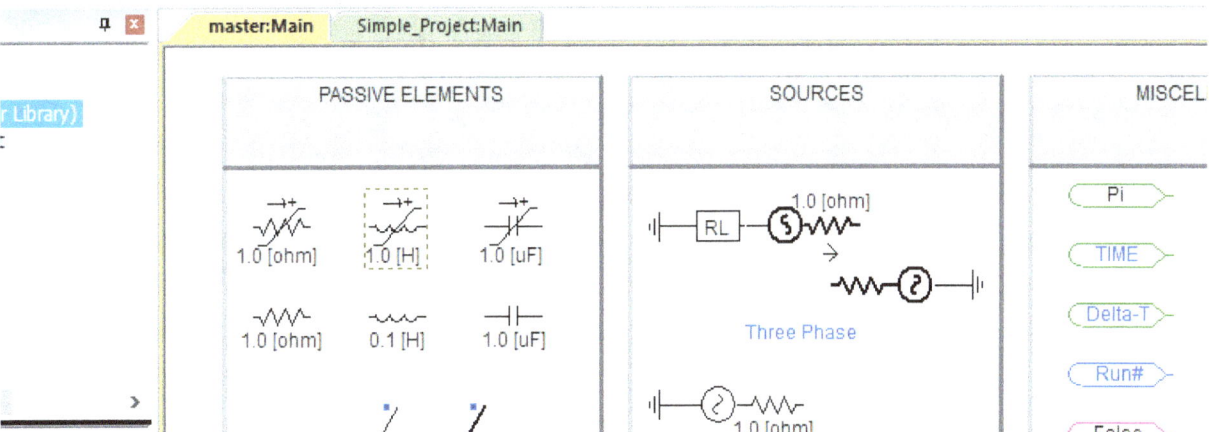

If we click at master:Main next to our Simple project we will see a lot of sections with elements. Those is where all the elements that are used in circuits inside the PSCAD, are stored. User can insert those elements by copying and pasting the element in the blank window. This will be described in the remaining chapters.

If we right click on the blank page, we see the following menus:

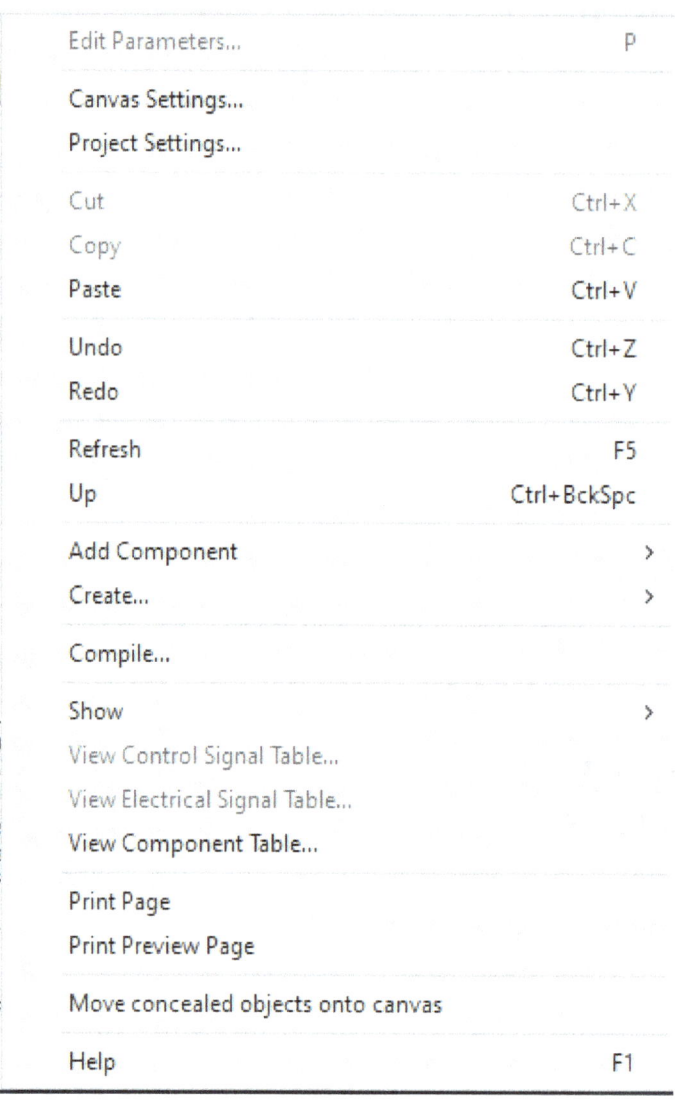

1.3.1. Canvas Settings

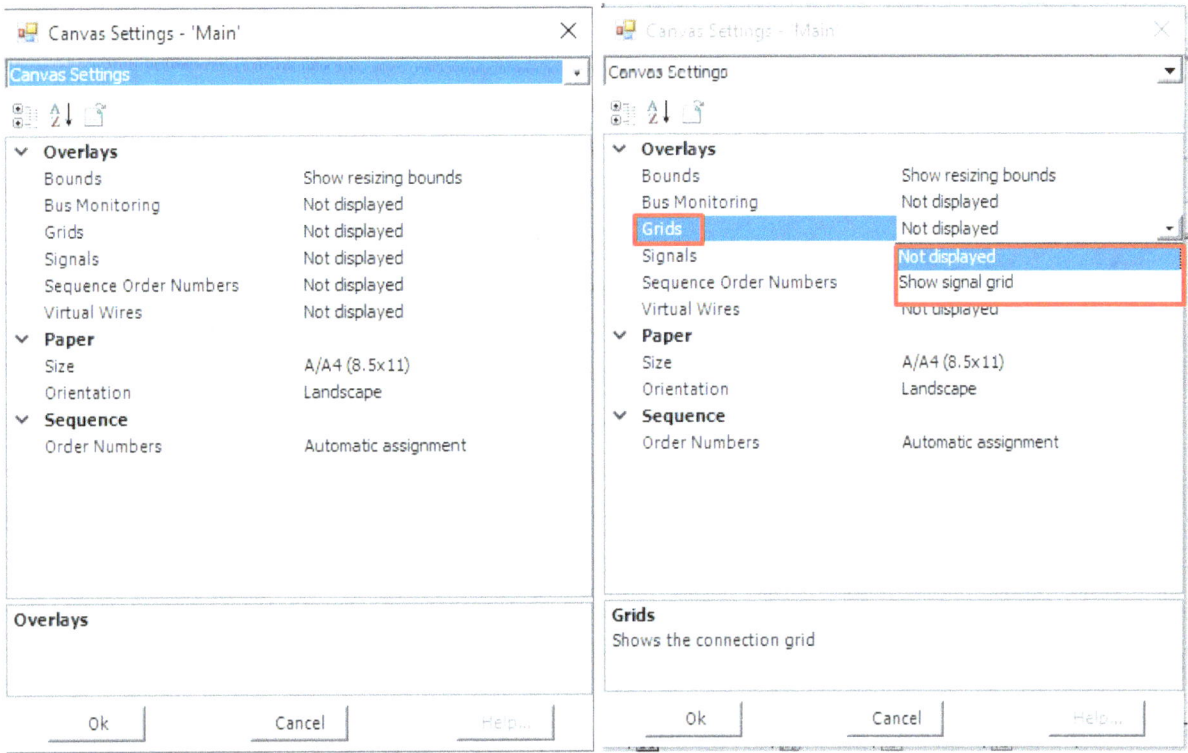

In the canvas settings the following settings are relevant:

- Overlays – It is an option that gives the user a more detailed view of the grid where he is working. Those settings consist on showing more data values, sequence numbers of element, routing, wires etc.
- Paper – User here can select the desired space format (A3, A4, A2 etc.) and orientation.
- Sequence – Usually left as default which means that elements are ordered automatically during dynamic modeling.

1.3.2. Project Settings

If we click project settings, we see the following menus:

- General

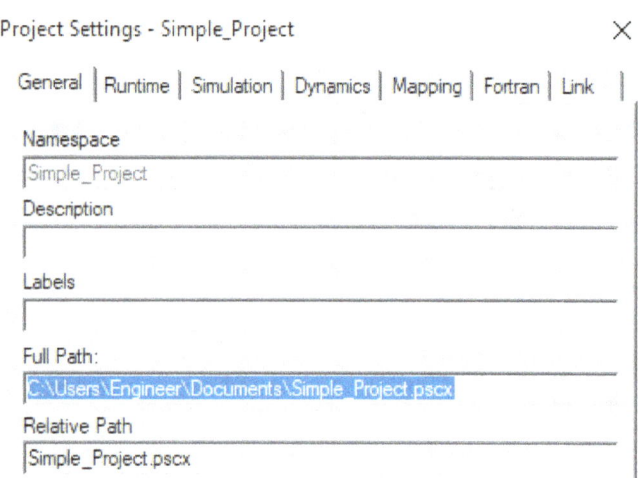

It is the main section where we can find some very basic details for our project and the software itself. Those settings vary from the name, path and description.

- Runtime

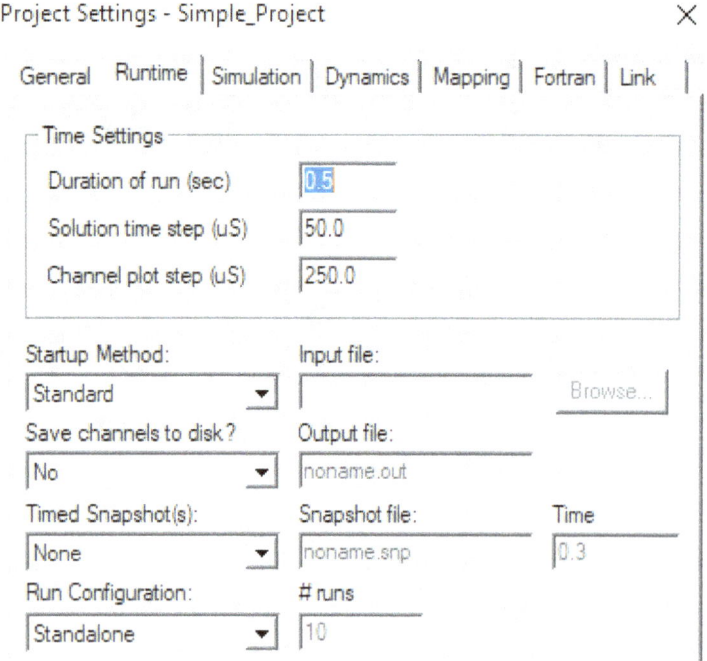

In the runtime user can set the time settings of the simulation. Those include the simulation time period, the time step, and the channel step plotting step size.

The rest of the settings usually are left as default and those settings include channel observe settings and snapshots.

- Simulation

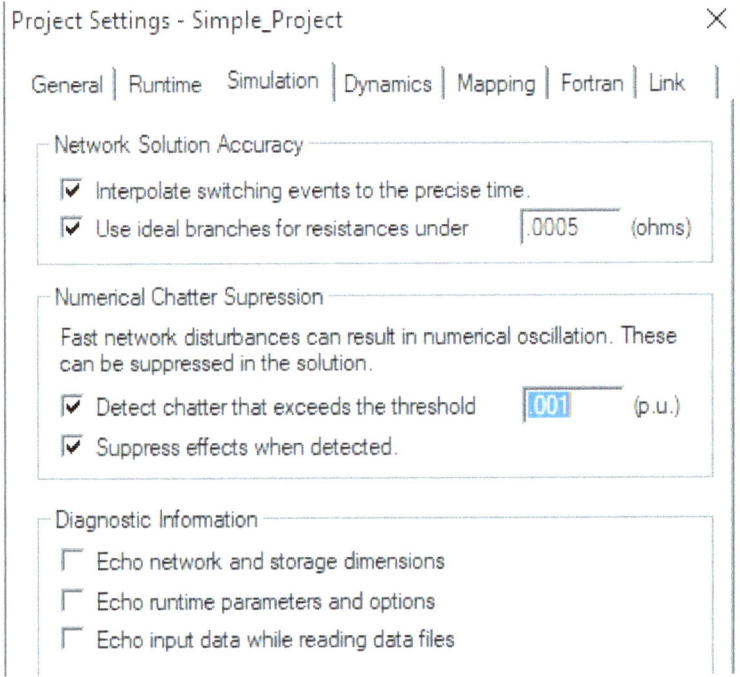

Simulation menu includes some relevant settings regarding the safety of simulation. User here can select network solution accuracy depending on the values of the elements that he is using during his project. Usually, all the settings here are left as default unless further modifications are required from the project.

Dynamics

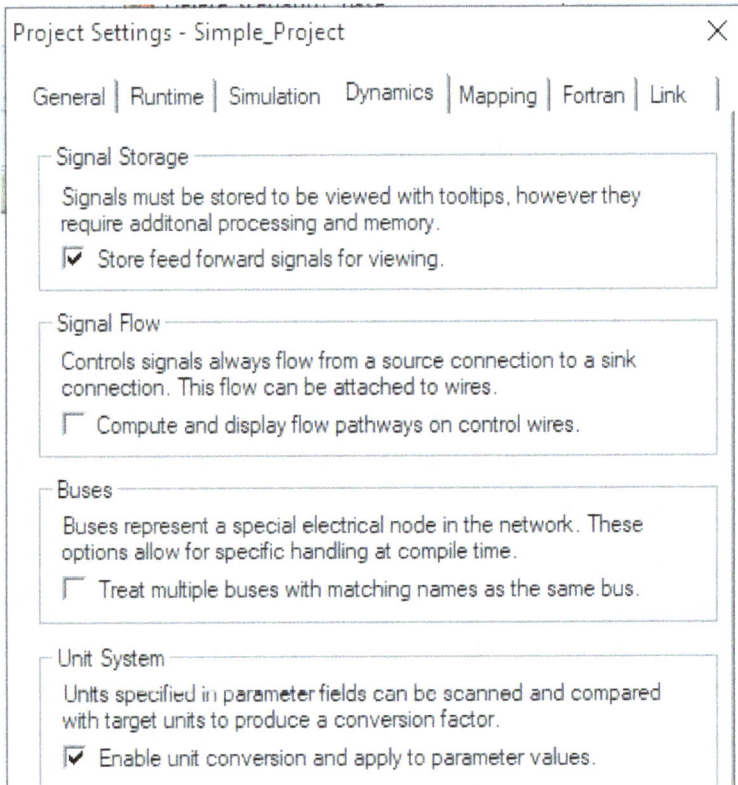

Dynamics include some settings that can simplify dynamic simulation. Those settings include storing signals, and configuring relevant settings for busses and unit systems.

- Mapping

Project Settings - Simple_Project   ✕

General | Runtime | Simulation | Dynamics | Mapping | Fortran | Link

Network Splitting

Networks can be split along T-Line boundaries to improve speed and use less memory.

☑ Split decoupled networks into separate matrices (recommended).

☐ Combine isolated non-switching local networks.

☑ Optimize memory allocation for decoupled network matrices. This option may require changes to user-definition script (recommended).

☐ Blend subsystems connected across module component boundaries.

Matrix Optimizations

Matrix solution will run faster or use less memory by optimizing the order in which elements are placed.

☑ Optimize node ordering to speed up solution.

☑ Move switching elements to speed up solution.

Dynamics include some settings that can simplify dynamic simulation. Those settings include storing signals, and configuring relevant settings for busses and unit systems.

- Fortran

Project Settings - Simple_Project   ✕

General | Runtime | Simulation | Dynamics | Mapping | Fortran | Link

Additional Source (.f) files

[   ] Browse...

Runtime Debugging

☐ Enable addition of runtime debugging information.

Checks

Enabling compiler checks may result in a decrease in simulation speed. It is recommended that these checks be used during project development stages only.

☐ Array & String Bounds   ☐ Argument Mismatch

☐ Floating Point Underflow   ☐ Uncalled Routines

☐ Integer Overflow   ☐ Uninitialized Variables

In this menu user can include fotran files as fortran is the base language of creating models and simulation in PSCAD.

- Link

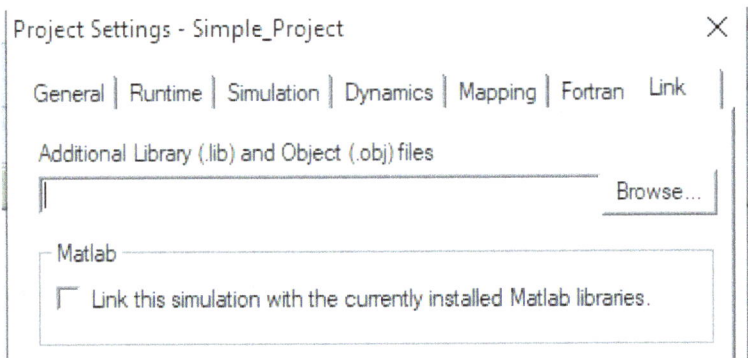

PSCAD also offers the option of linking additional libraries of models created by the users and also interfacing with Matlab.

1.4. Creating a simple circuit

Now we will begin to create circuit in PSCAD. To insert elements on the blank grid the user has the following options.

- By Right Click and select element

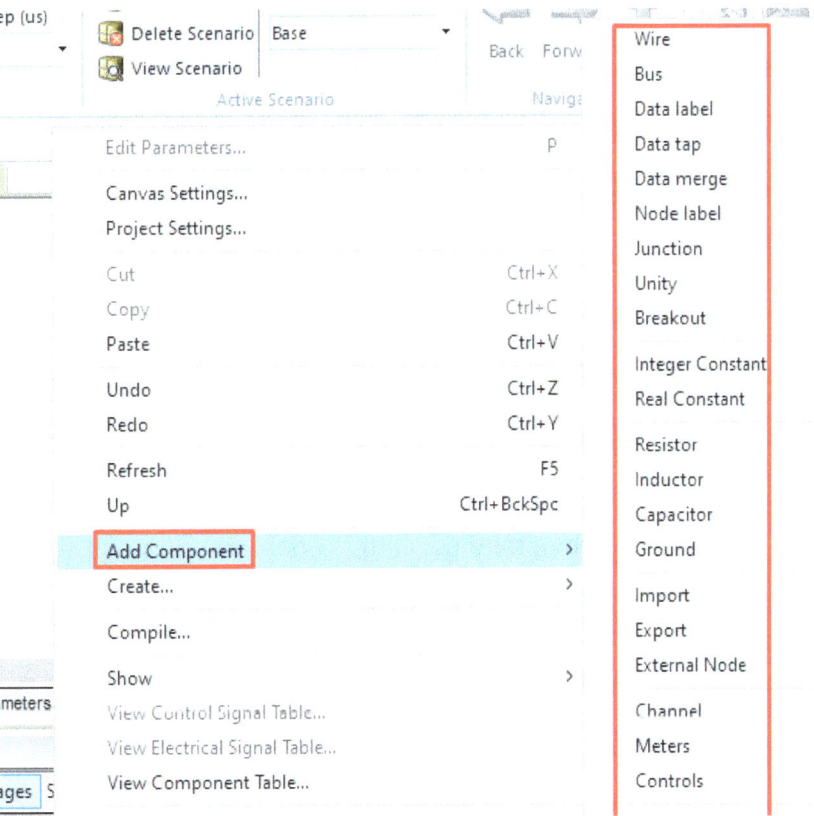

- By selecting components in the main toolbar menu:

If we click models in the toolbar menu and select any categories and the elements inside category, we can see that the mouse automatically makes possible the view of the element to be placed in the

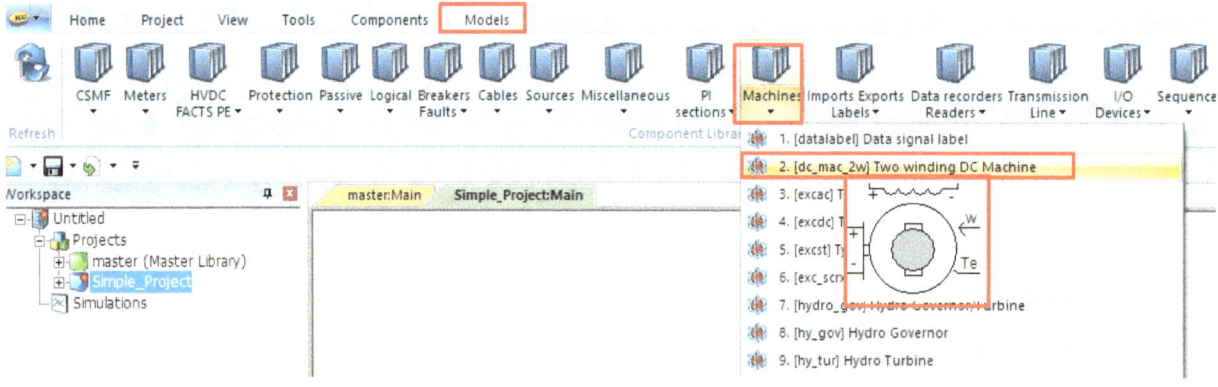

- By master main library

PSCAD has a built-in library for all the elements that we need. This library can be updated or changed. To select and element and place inside the schematic diagram, simply click copy and paste into the diagram.

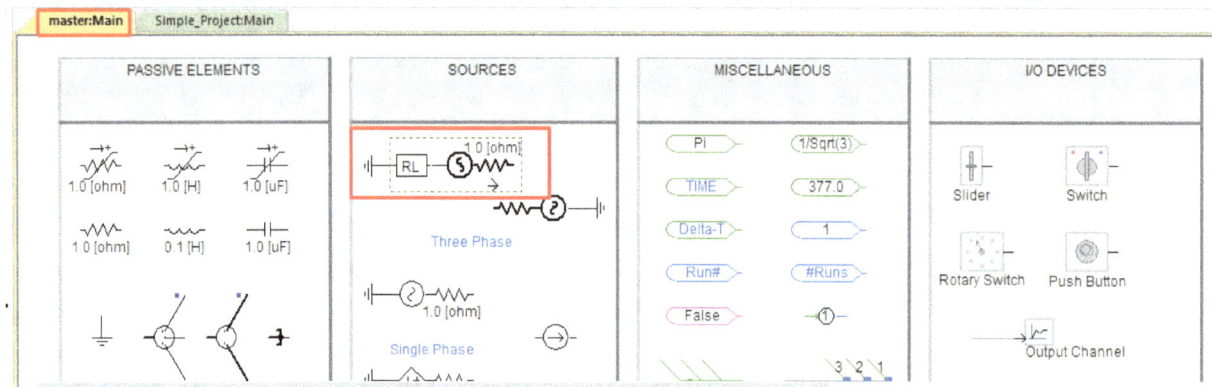

User is free to select any of the above method. In this book we will go for the third method as it is easier to use and all the selected elements are categorized and well explained.

To enter inside the elements, click on the big arrow on the elements category:

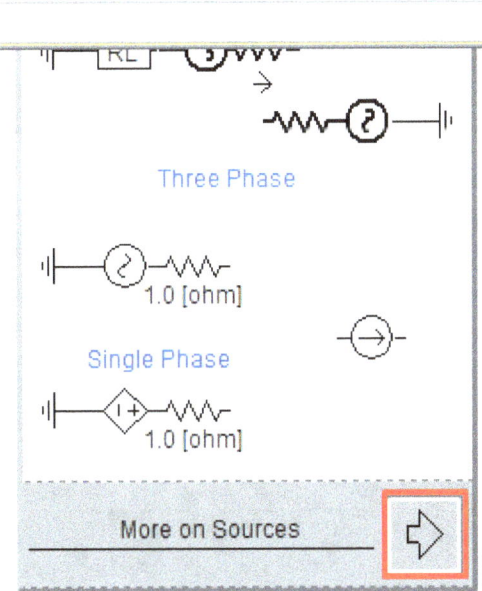

From the sources select the component as in figure below and copy.

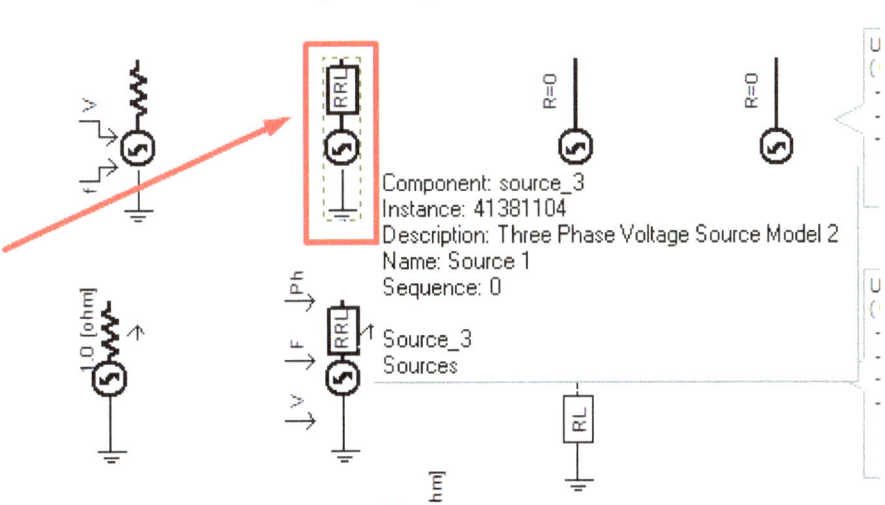

Then head to the main schematic diagram and paste it like below. By selecting the element and clicking it, type R to rotate the element as desired:

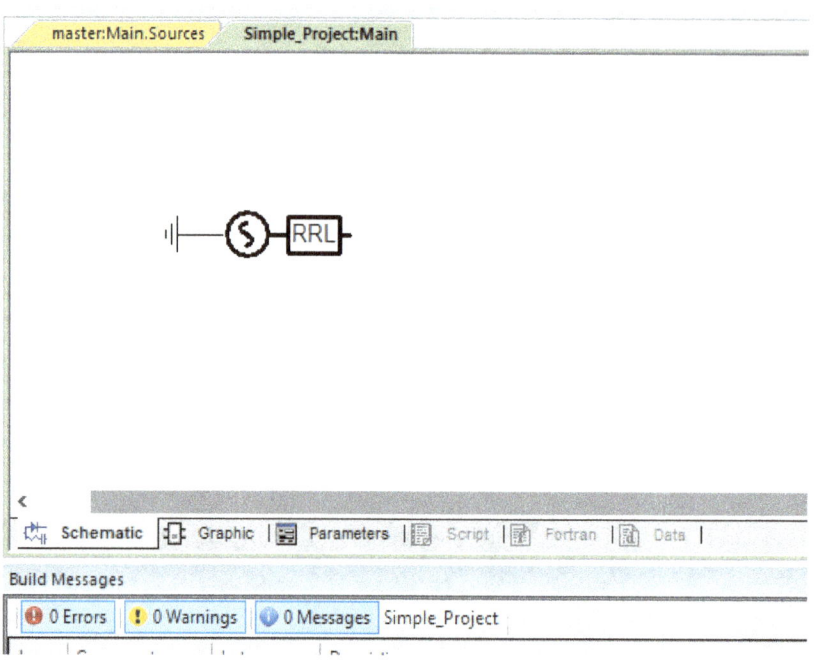

To edit the element double click and let's configure some main parameters:

In the main configuration menu change the element as an Ideal one with R=0

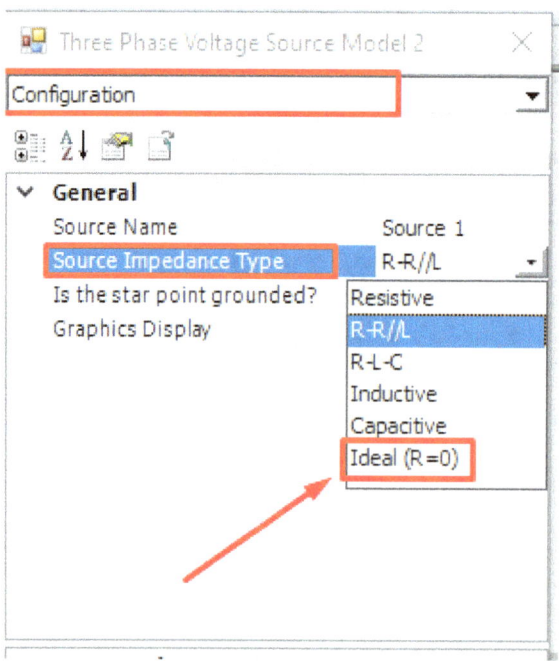

Also, to have a clearer view for all the elements that are more than one phase, select the 3-phase view in the Graphic Display.

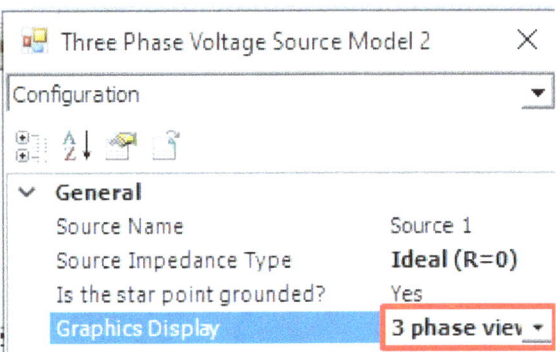

This will result in the following view:

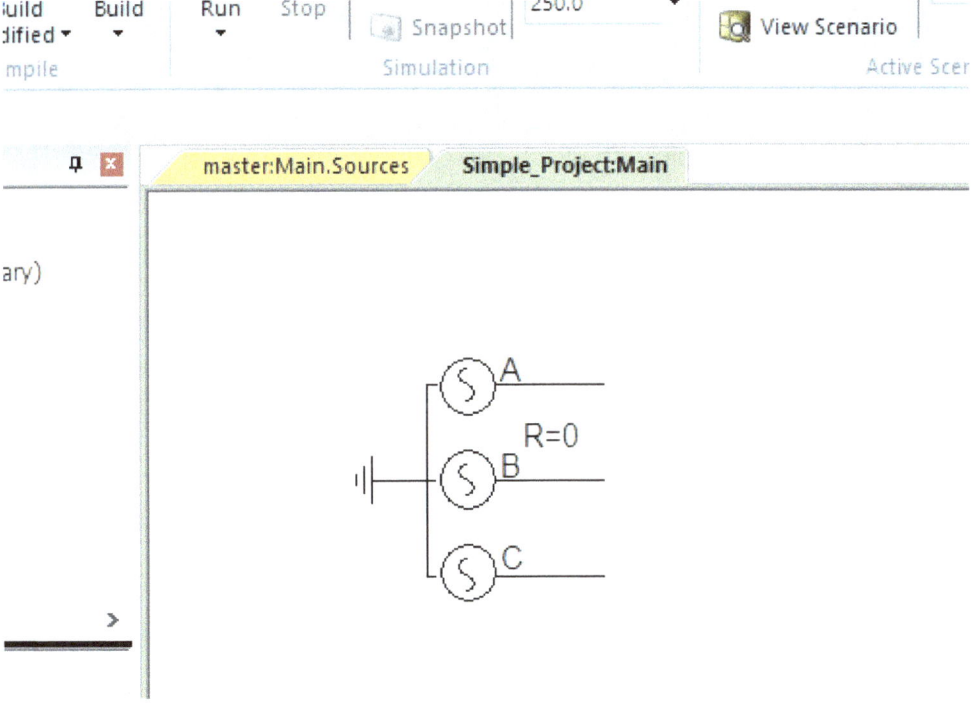

Now we will connect 3 filters to the circuit. Those filters will be passive elements and respectively R, L and C configurations. To search for filters, head to passive category on the right and select the three series RLC tuned filters. Click the filter, copy it and paste 3 time for the three respective phases.

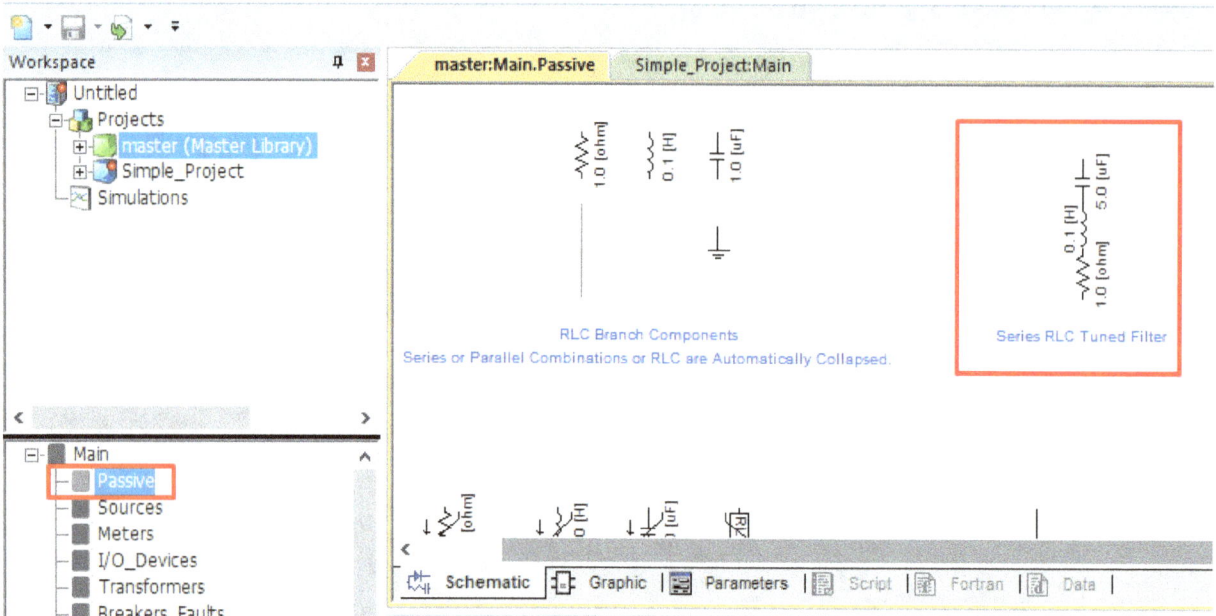

Rotate and adjust to have the following view:

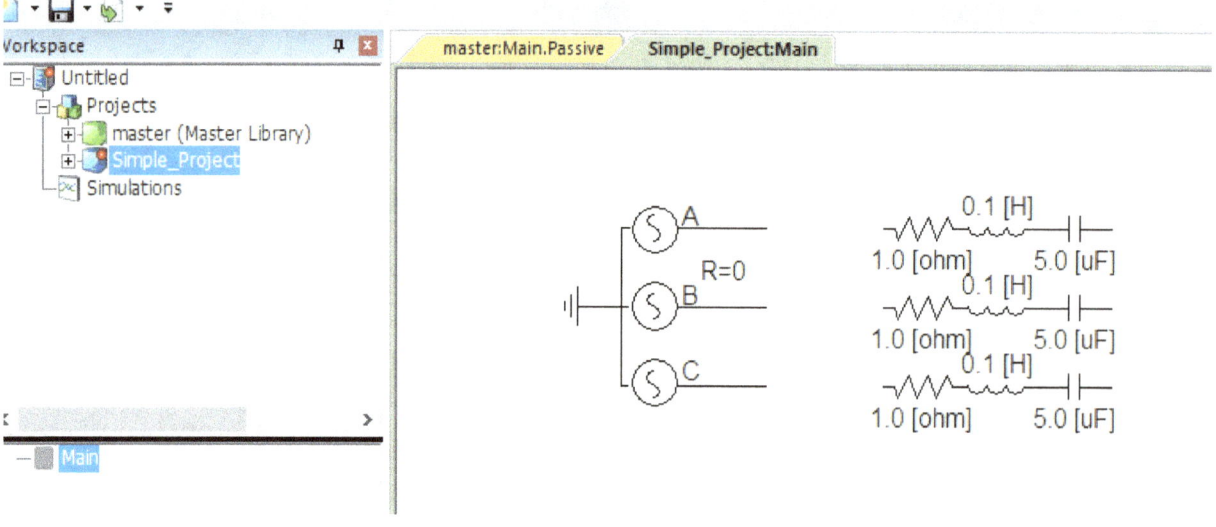

To close the first circuit, we need to connect elements, add some measurement devices and a grounding point. To connect the elements, activate the wire mode and connect elements with the wire tool. The wire mode can be activated by selecting wire mode in the components toolbar or by typing Ctrl + W:

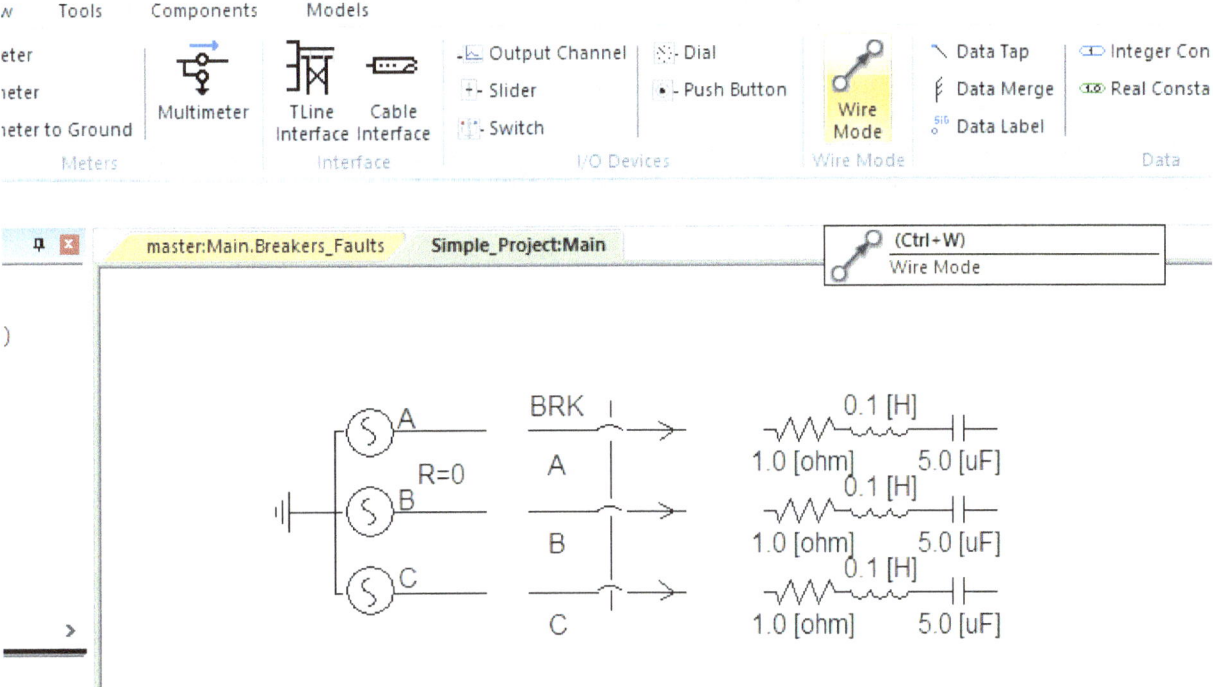

Then click the wire between two ends of elements. After connecting elements, the view should look like this:

To insert metering devices, user can insert those devices by selecting the meter at the component menu or selecting more complex metering devices in the meter's library:

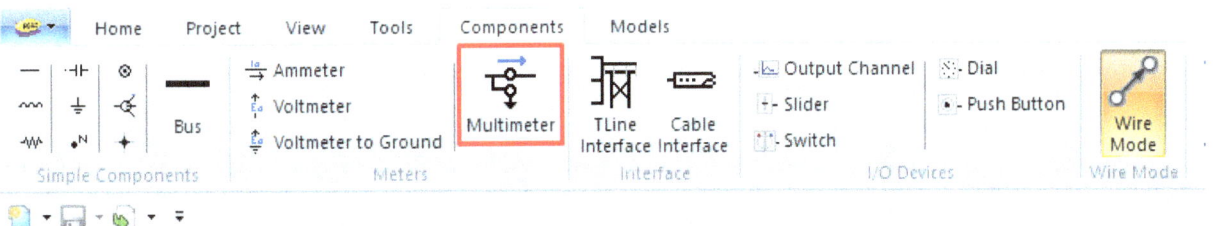

Click the multimeter as it is shown in the picture above and insert 3 multimeters for the three phases respectively like in the figure below and connect with the phases through the wire:

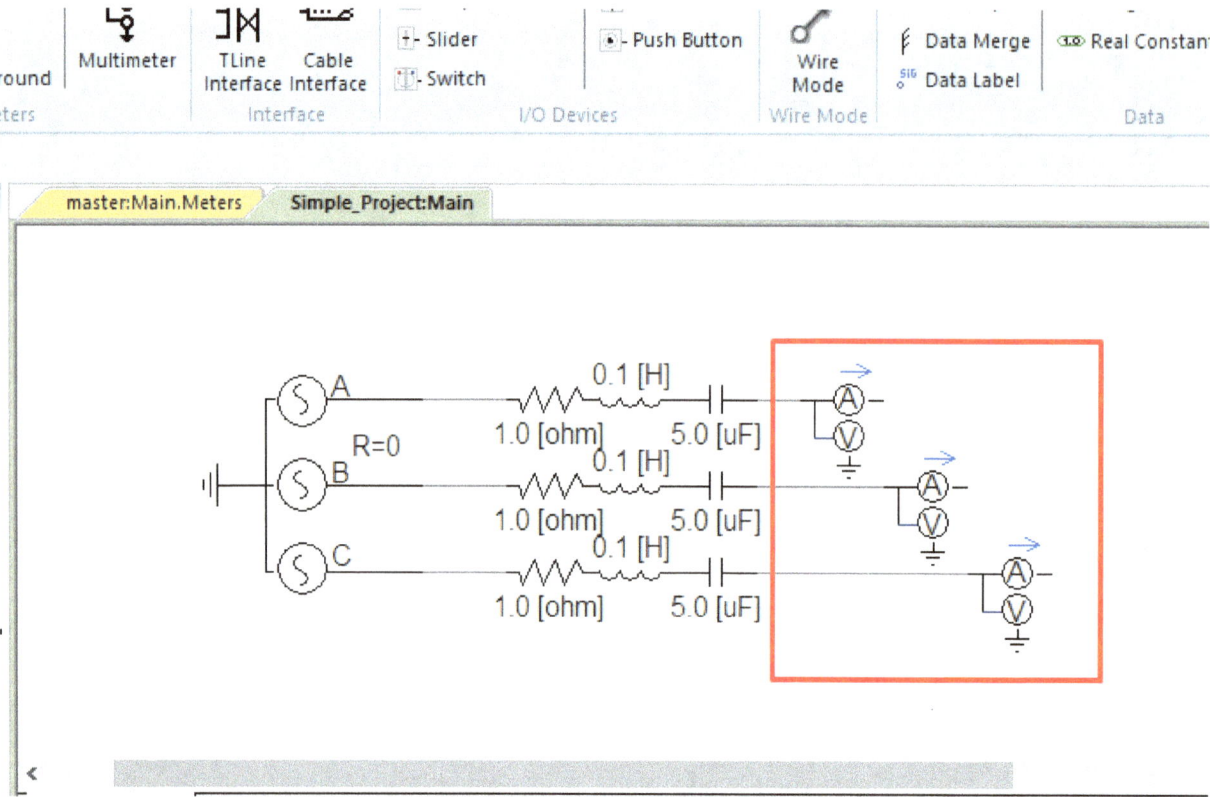

Now let's close the circuit by adding a grounding element like in the figure below:

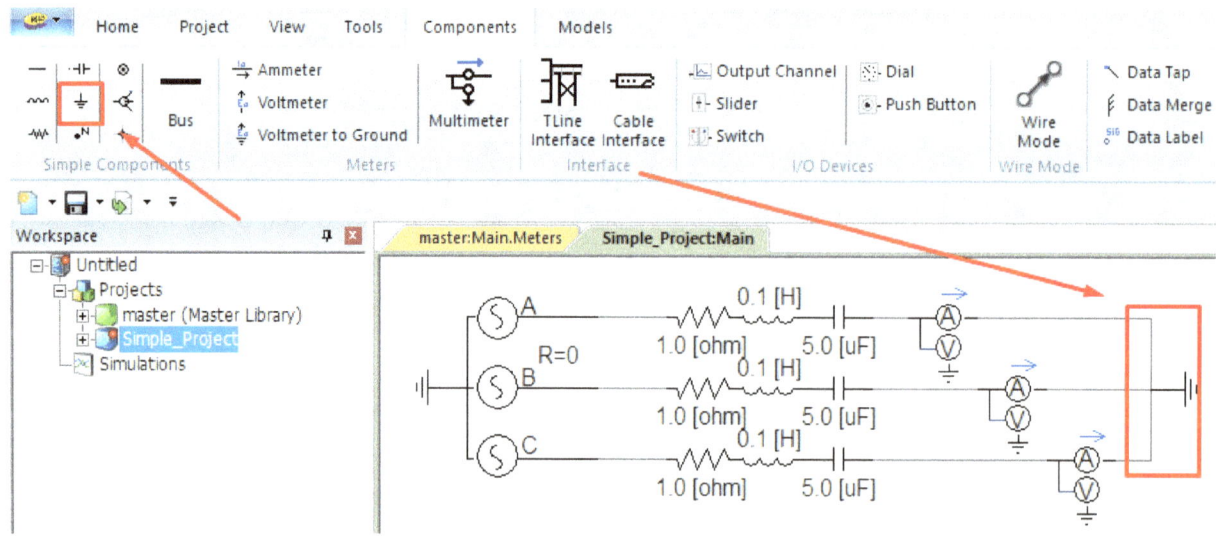

1.5.   Outputting channels

Now that we completed the first circuit let's learn how to output channels or variables. Every variable or channel in PSCAD can be routed and observed by its name. Let's double click one of the measurement devices.

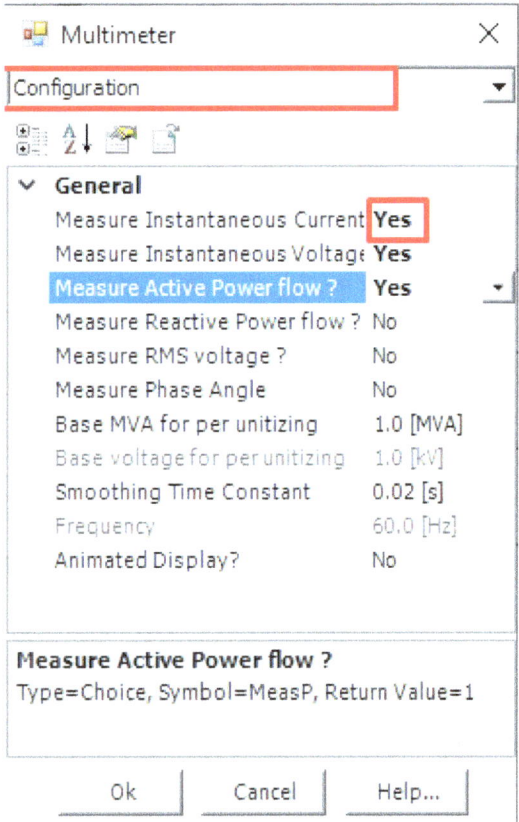

In the configuration menu choose which variables that are available to be measured by changing from No to Yes.

Then the next step is to set a name to those signals and to route them. Let's head to the next menu of the Signal Names.

Every signal accepts a text type so user is free to set whatever name or text to identify the signals. In our case we will measure Voltage, Current and Power and we will call them specifically by phase names like Ia, Ua and Pa.

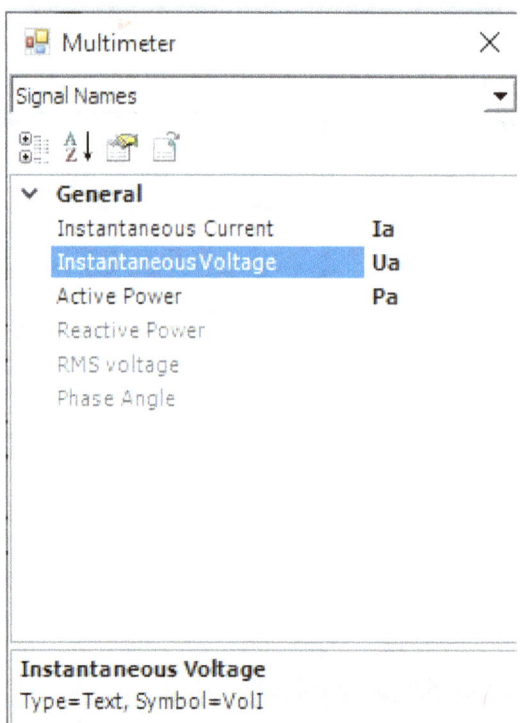

Configure the other multimeters in the same method like the multimeter above for the other 2 phases with respective labels of voltage, current and active power.

1.5.1. Routing signals

To route signals in PSCAD we need the so-called data labels. The data labels used to route signals by acting like taggers inside the PSCAD environment. Under the components menu, find the Data Label icon and insert into the diagram as below:

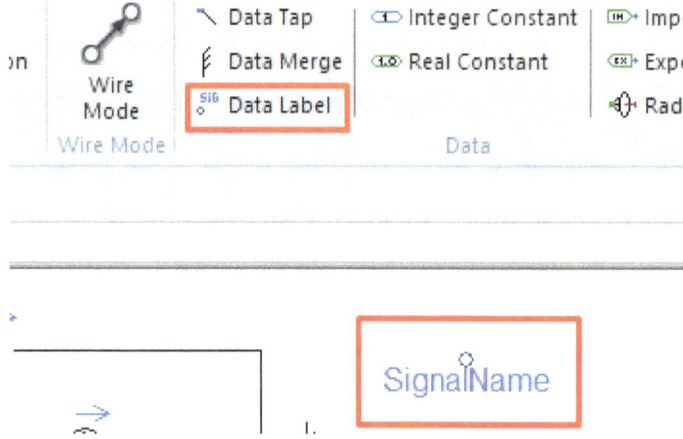

Rename the SignalName label as the name of the signals that we set in the previous step. As we have 3 variables for the three phases, we need 9 labels in total. Insert 8 more labels and rename them respectively to the names:

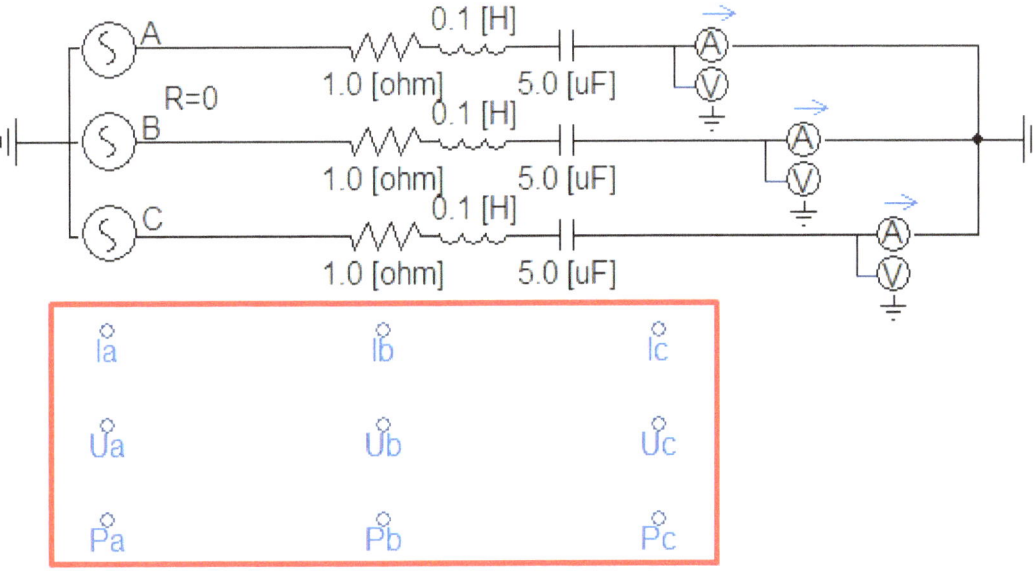

Now that we have successfully routed our signals, we have to connect the signals to output channels. From the component menu select the Output Channel and connect each signal name with the output channel.

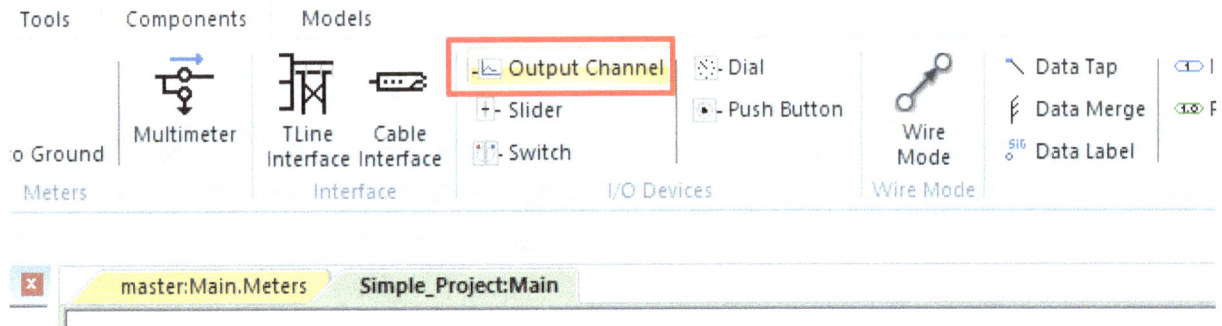

The diagram should look like below:

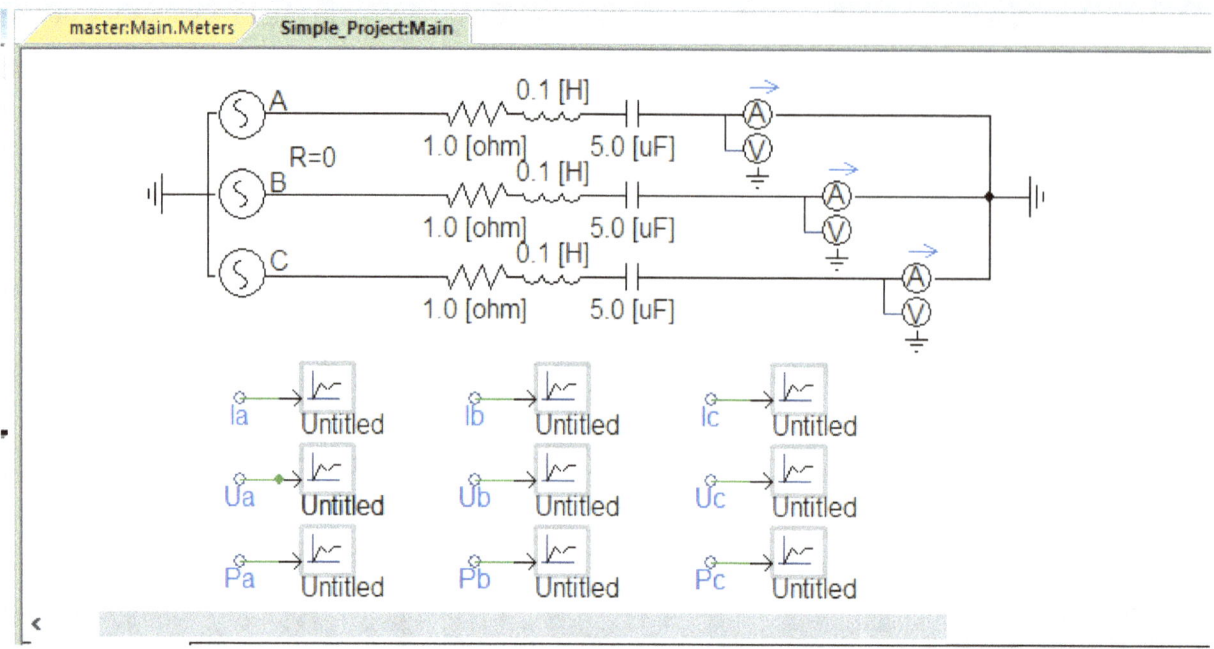

Now we can create plots in the following ways in PSCAD:

- By inserting a Graph Pane and adding signals
- By extracting the plot direct from the output channel

Head up to the component menu and select Graph Pane and insert it into diagram:

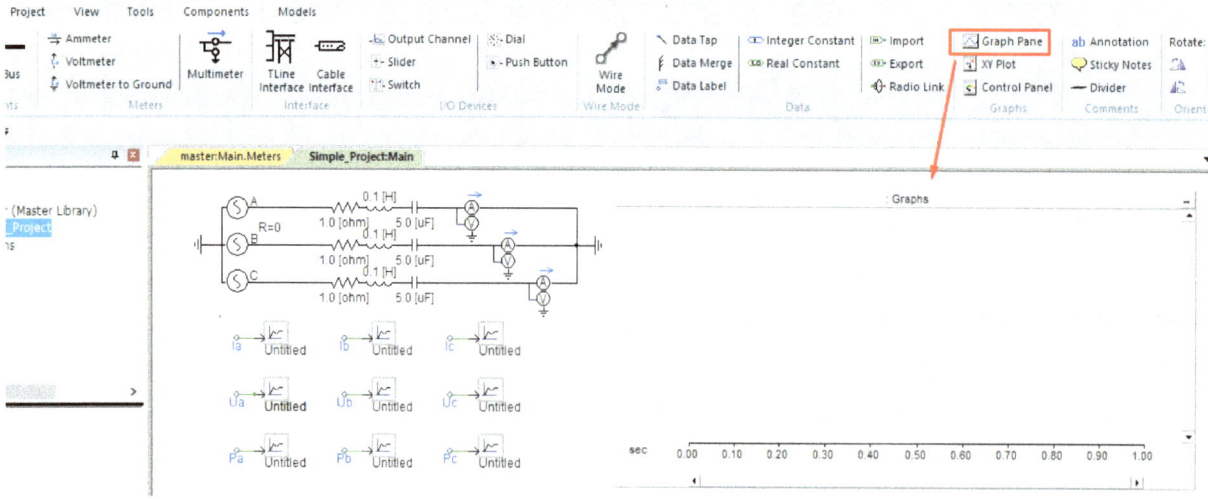

To insert individual plots, right click the channel, go to the Graphs/Meters/Controls, select "Add as curve" and head to the plot and paste the curve into the Graph Pane:

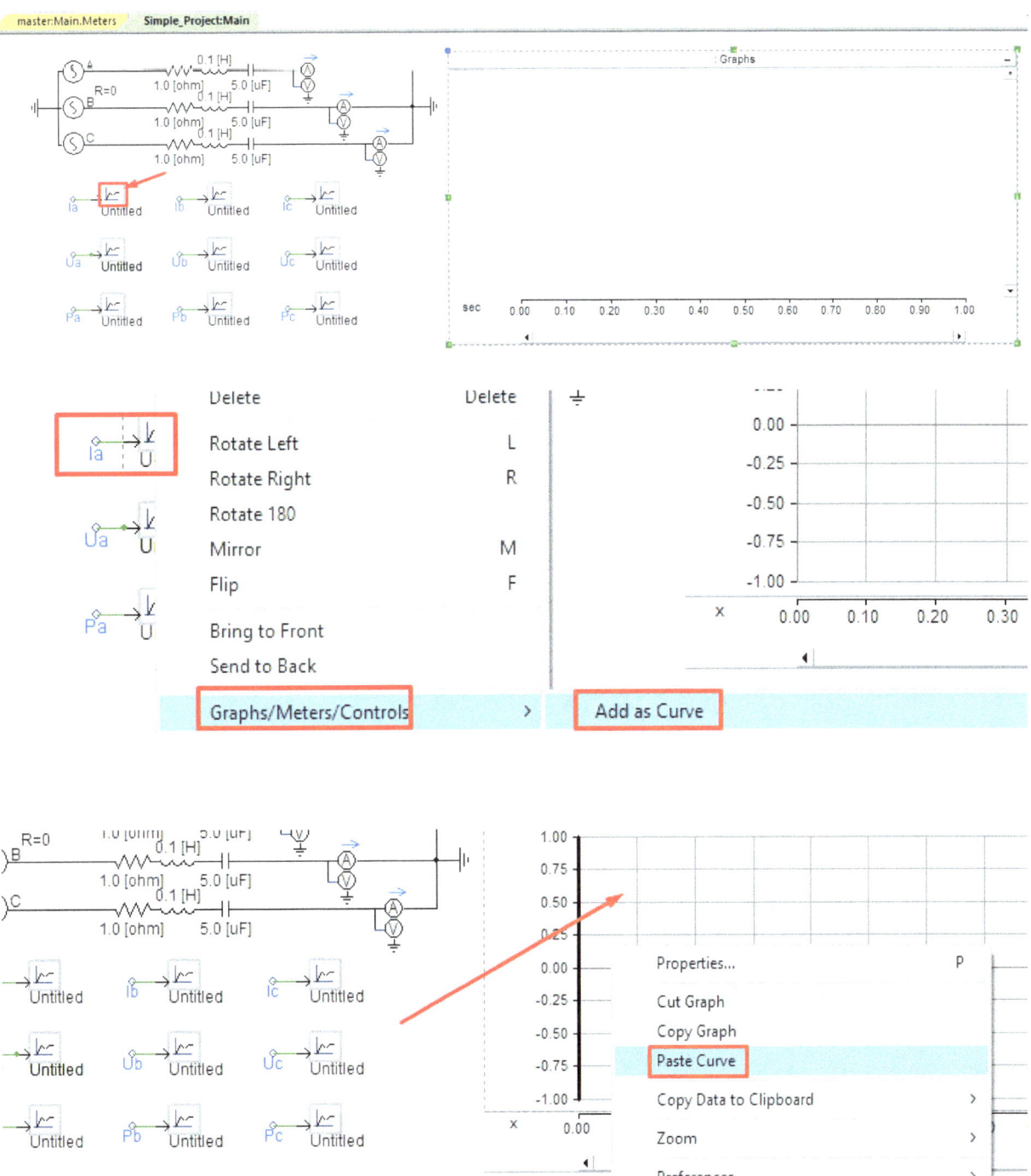

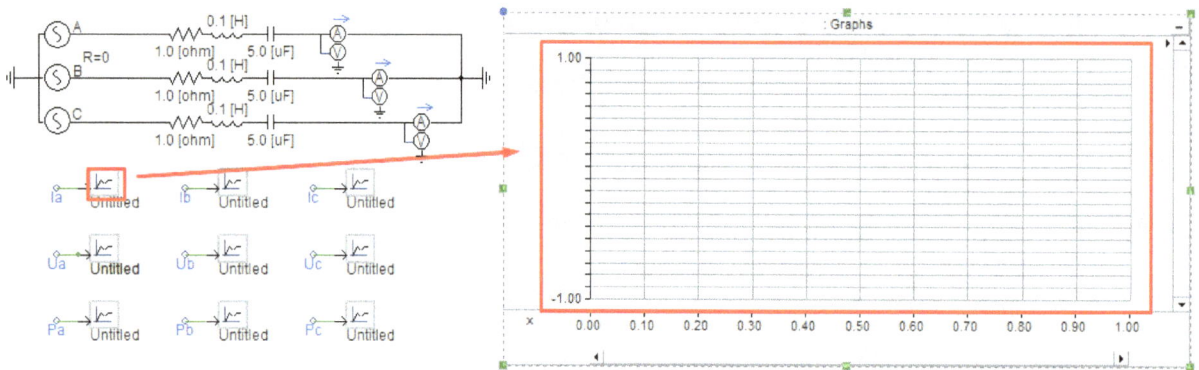

As we can see below, we can see the "Untitled" created which means that the signal is logged into the plot.

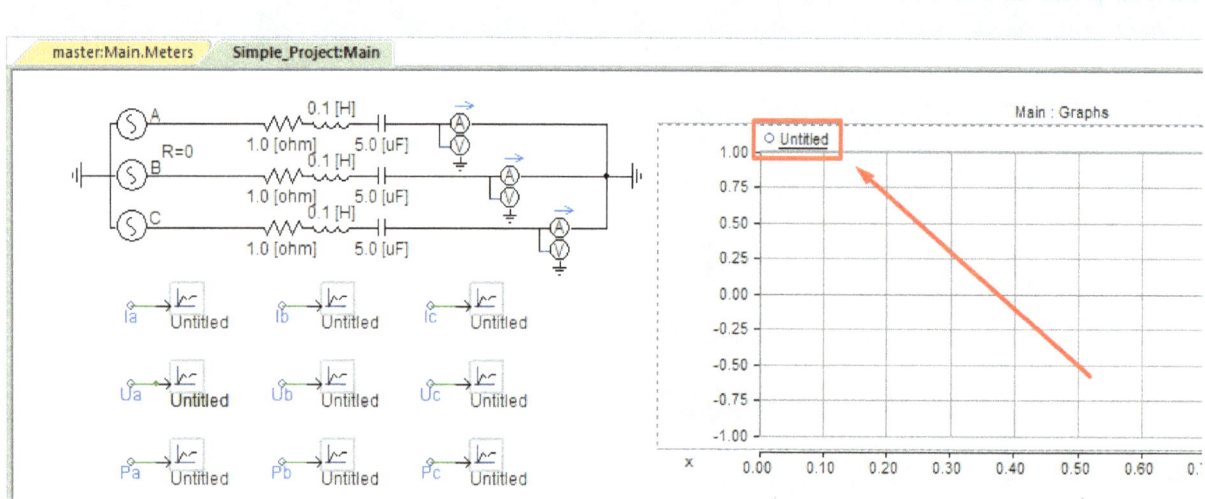

For the next method the user can generate any kind of plot directly from the measurement block. To perform this, right click like in the figure below, head to Graphs/Meters/Control and Click Add Overlay Graph with signal. This will create an individual plot of the signal.

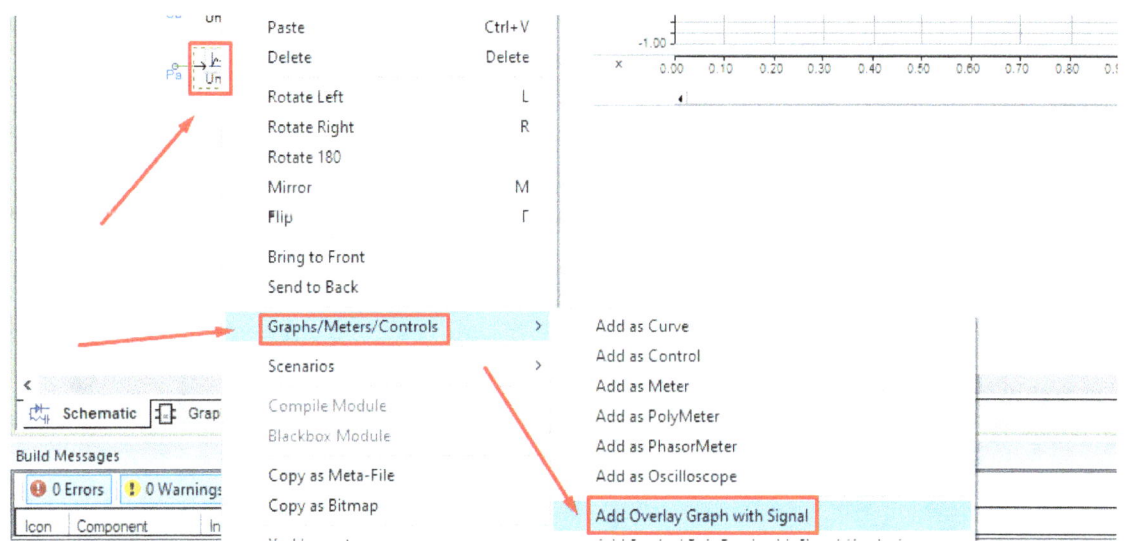

As we see a new plot is created.

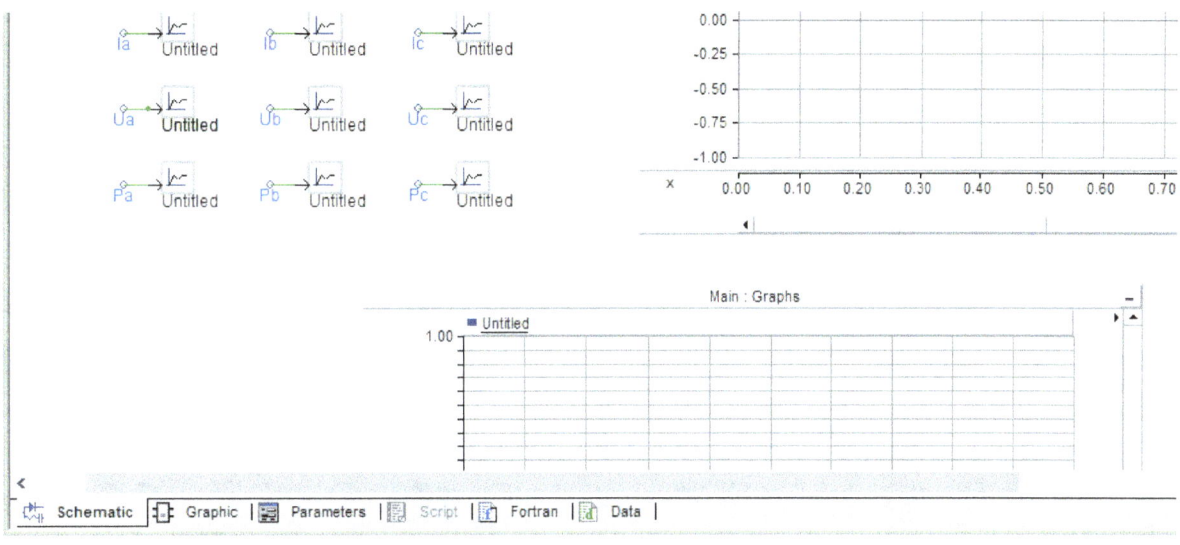

1.6. Run a Simulation

Whatever type of system that we want to simulate, the steps to perform a simulation in PSCAD are the same. The first step is making sure that all configurations of elements are done correctly and those elements are connected correctly to each other.

- Building the model

The first step consists on building the whole project and preparing the simulation. This can be performed by heading to home menu and click the "Build" icon:

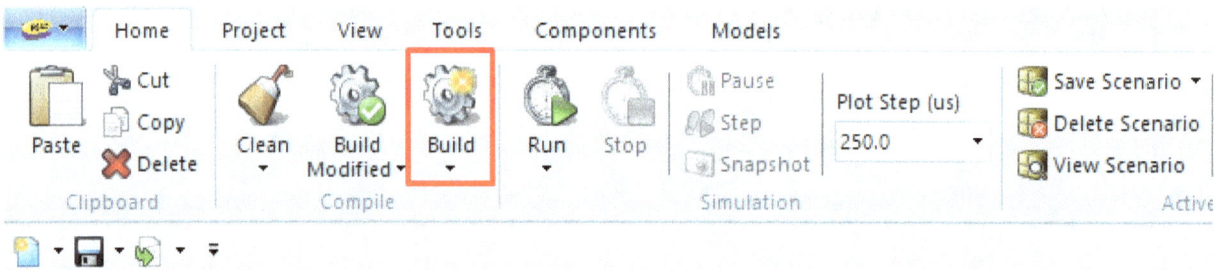

After clicking the button, the user should see in the bottom right corner two working mechanisms, which mean that the build is working and if no errors are shown in the build messages, then the build is performed successfully.

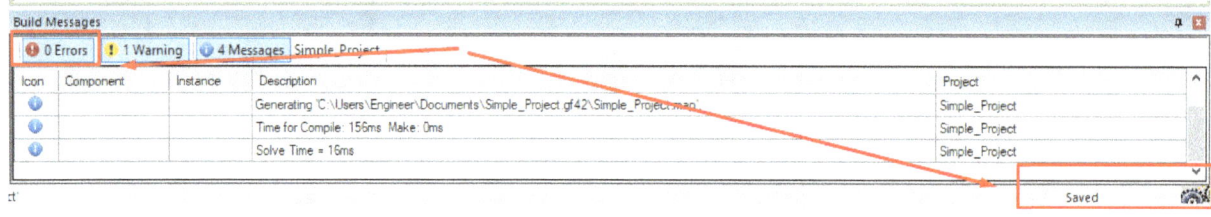

- Set simulation parameters

To set simulation time and step size of both simulation and plotting right click on the project diagram and select Project Settings

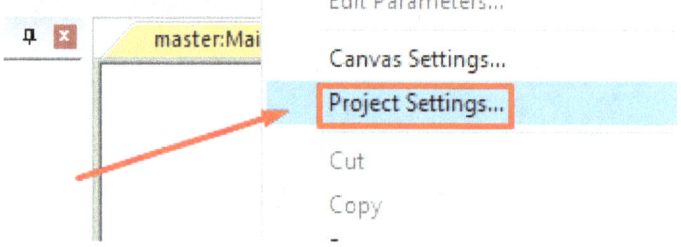

As explained in the beginning of this book the user can configure simulation parameters in the window below. In our example we will go with the default parameters of step size configurations and with a simulation time of 10 seconds:

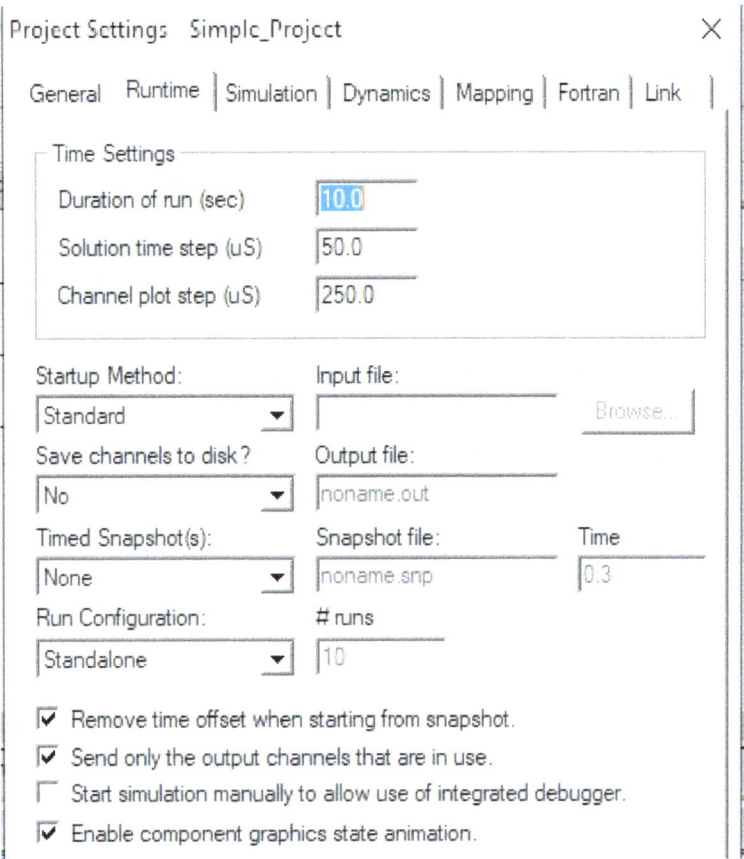

The user also can configure the simulation settings on the project toolbar like below:

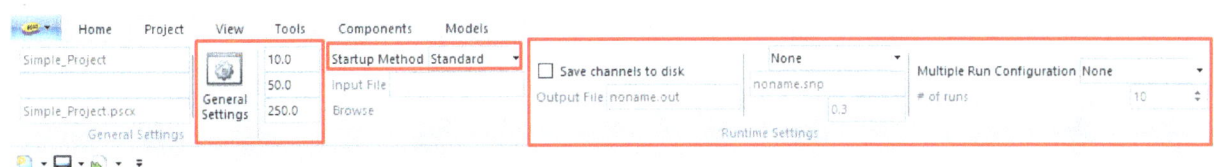

- Run the simulation

The final step consists on performing the simulation. To perform the simulation click Run in the Home toolbar.

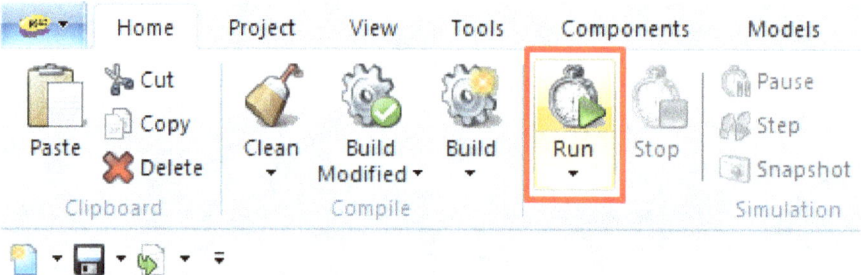

We can see the simulation details while running again in the bottom right corner like below:

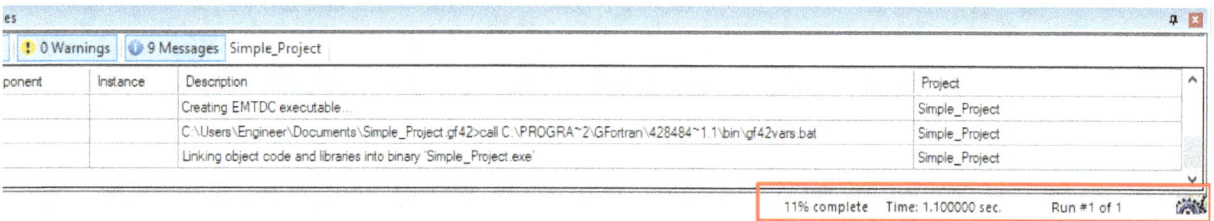

Now we can see the simulation is done and we can see the finished plot.

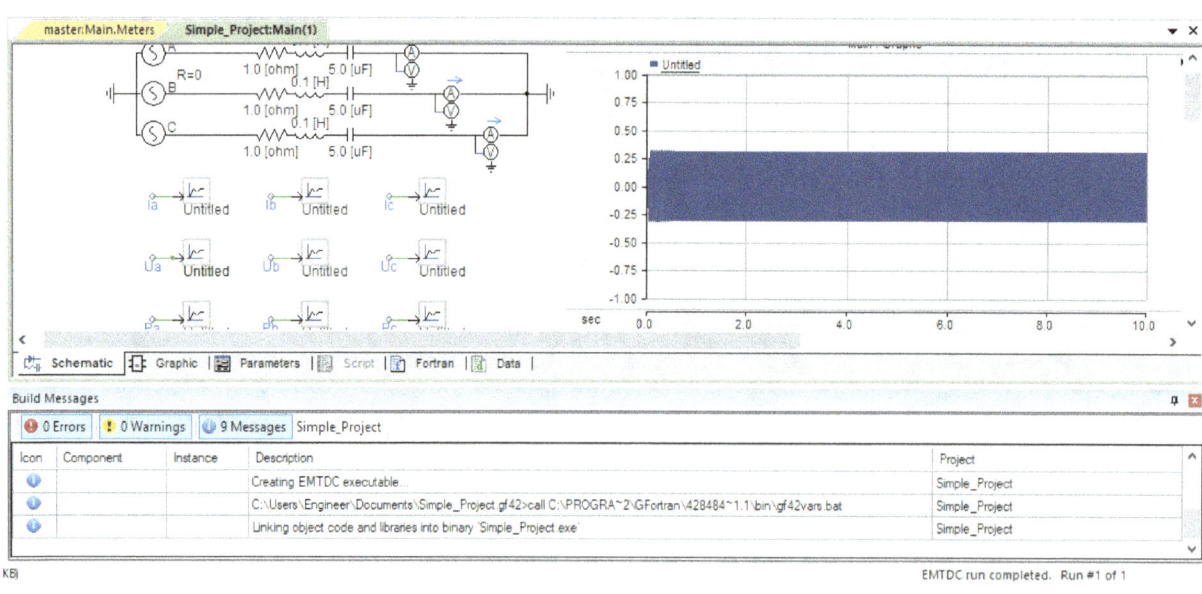

1.7.   Switchers & Breakers

As we know that PSCAD is one of the most powerful tools regarding Electromagnetic Transients simulations, breakers and fault is also a very interesting topic when performing very basic calculations to more complex studies. To insert a breaker into our PSCAD diagram head to model's directory and select Breakers & Faults and then select 3 Phase Breaker as shown in the photo below:

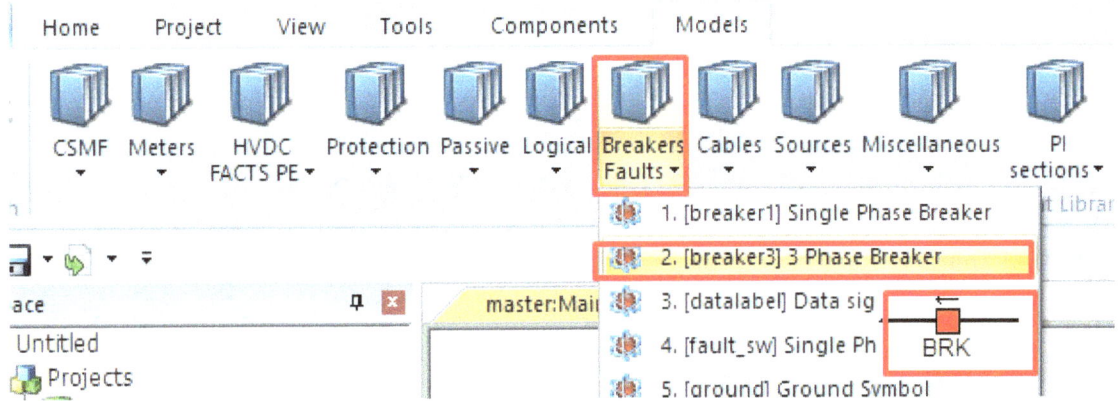

Then double click on the breaker inserted and select the three-phase view in the configuration.

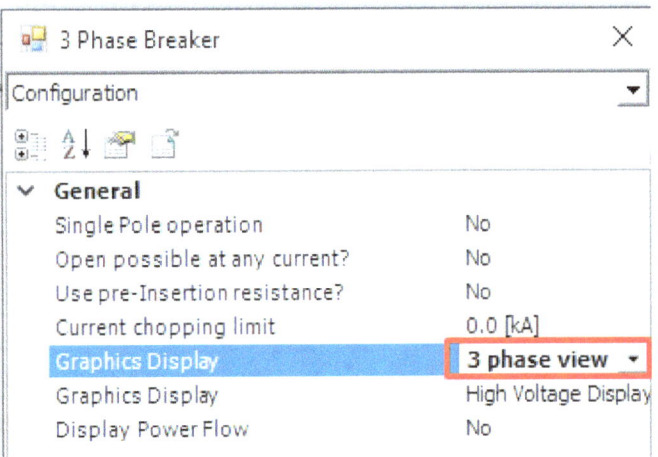

Then connect the breaker between the three loads and the power source by moving elements and using the wire tool (Ctrl + W). Then the graph should look like below:

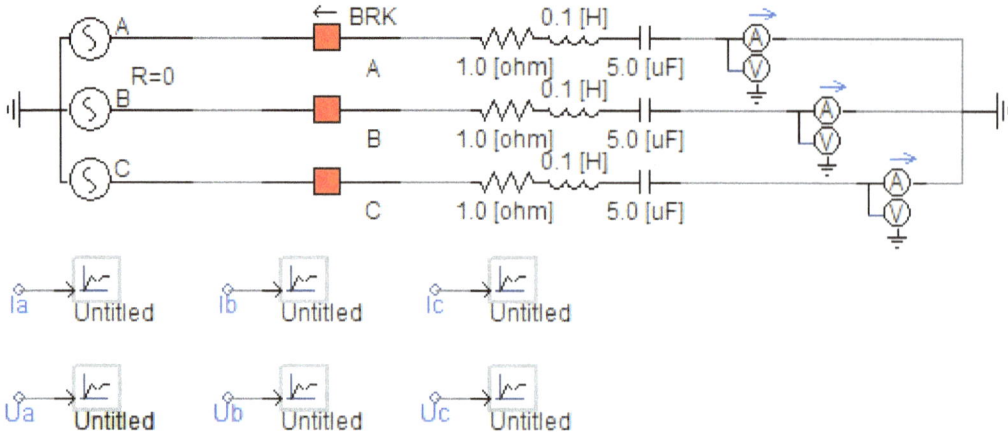

Now let's configure all elements and perform the simulation of a breaker.

Configure the source as below:

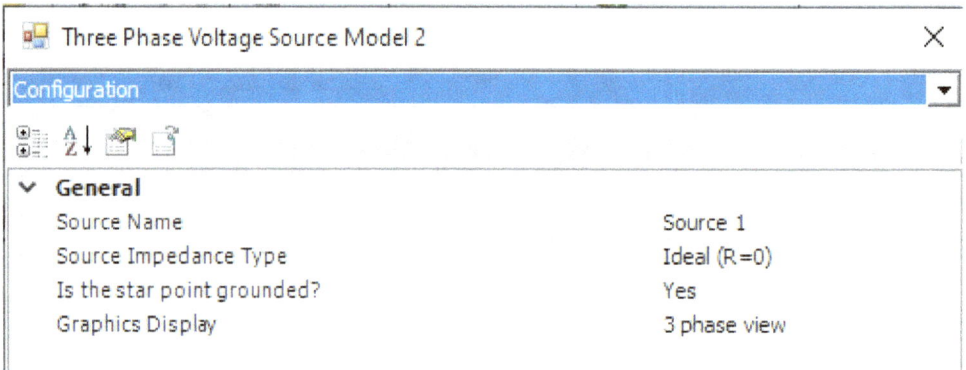

And with a voltage level of:

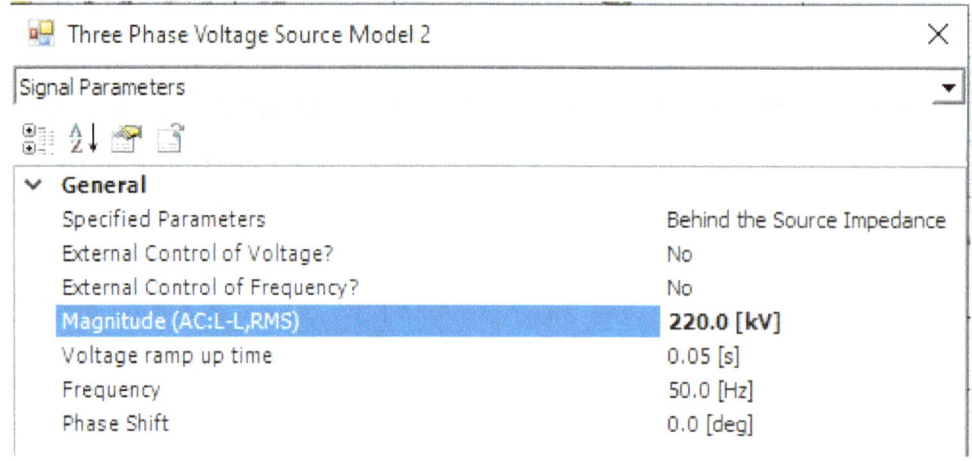

Regarding the loads we will go for the default values. User is free to change any desired value. Now let's configure the circuit breaker. Double click on the circuit breaker and head to the breaker main data. Before doing this let's explain how breaker can be configured in PSCAD. The two main scenarios are:

- Control the breaker like a switch
- Configure the breaker by its current limit.

Now we are going to configure the breaker as switch and we will control it during the simulation. In the Breaker Main Data change the name as "Breaker_1" as this will be used as our tag to control the 3-phase breaker.

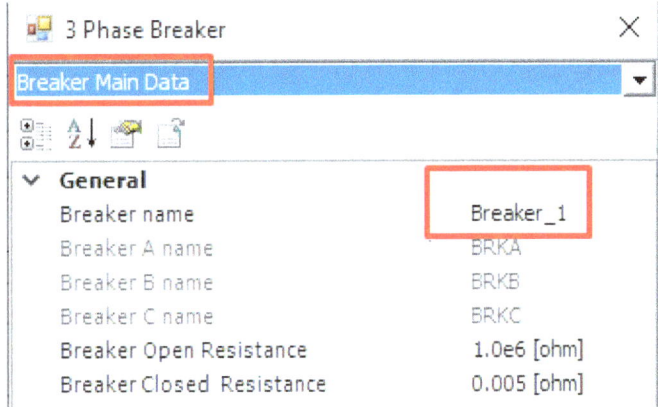

Next head to the Models, Breakers Faults and insert a Timed Breaker Logic.

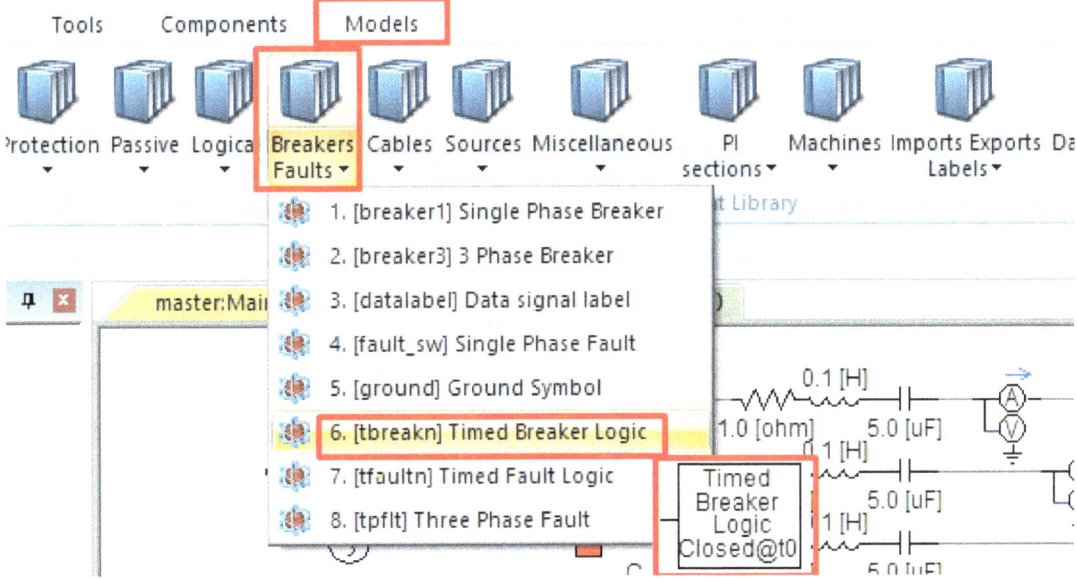

Now in order to connect our breaker with our control block we need a data label the same that we performed with the measurement devices. Let's insert a Data Label like in the figure below:

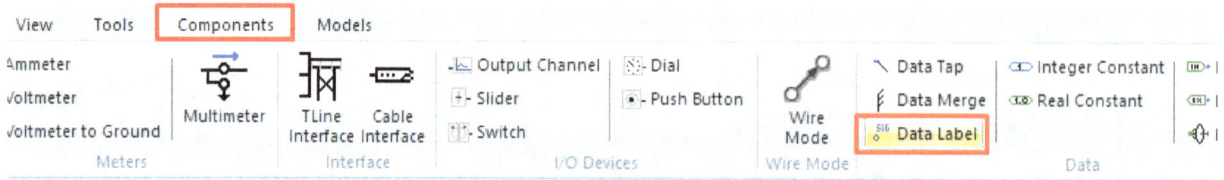

Rename the Data Label as the breaker name "Breaker_1"

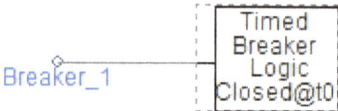

Let's configure the Timed Breaker Logic

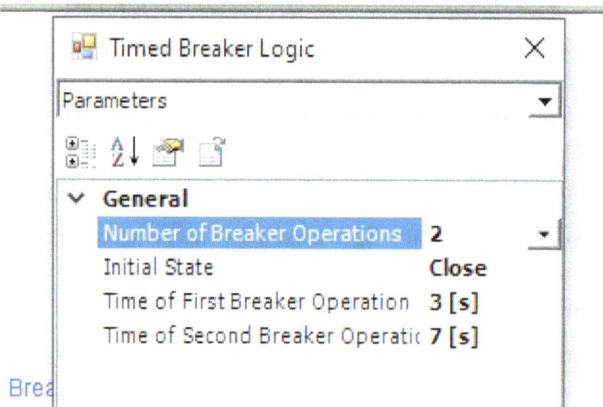

So, configure the breaker as in the figure above and let's explain the settings. The number of operations, which can be 1 or 2 means how many actions the breaker is going to perform. Initial state is set to "Close" which means that the next operations will be "Opening" and "Closing" again. If we would have configured the Initial State as "Open" then the rest of the actions would e "Close" and "Open" again. Then we set the time of the first action (Open) in 3 seconds and the next operation (Close) in 7 seconds. Next let's create 3 plots showing the currents of the three phases and let's perform a simulation.

After we run the simulation for 10 seconds, we should have the results like below:

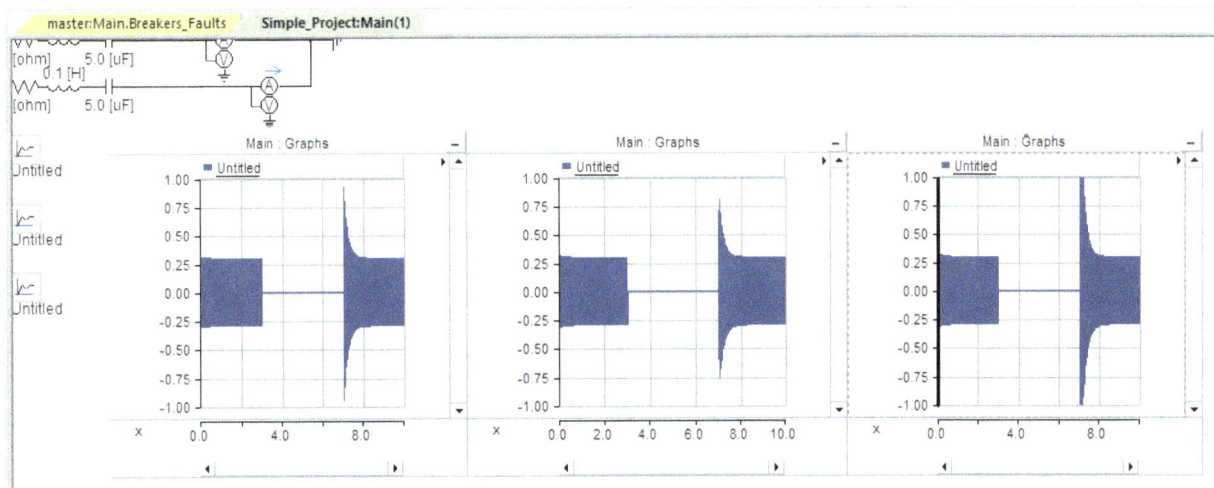

1.8. Fault simulation

Let's simulate a fault in our diagram. To create a fault, head to Breakers and Faults and insert a Three Phase Fault.

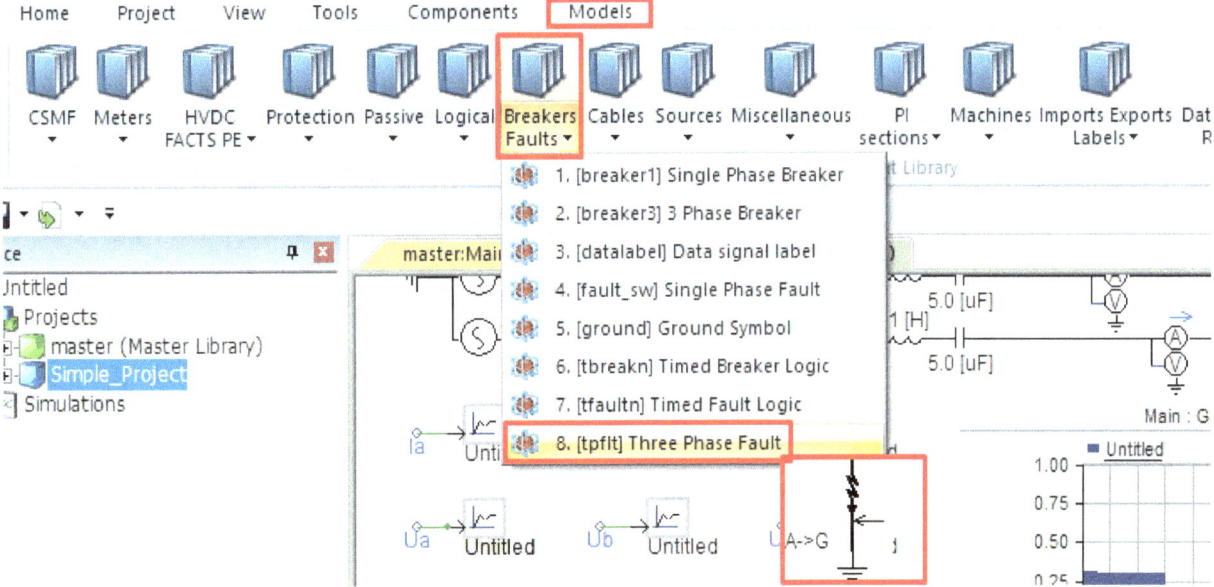

Convert the element to have the three-phase view in the configuration menu and connect to the three lines like in the figure below:

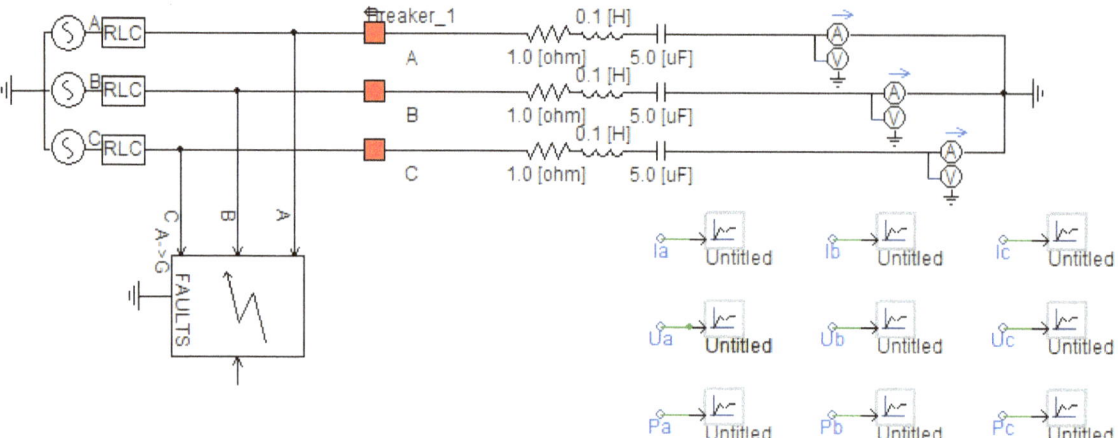

Before configuring the fault let's configure the three-phase power source. As for the fault, we will configure a Source impedance type of R-L-C instead of an ideal one. In the ideal one the short circuit power is almost infinite and we cannot see the fault short circuit power we will apply.

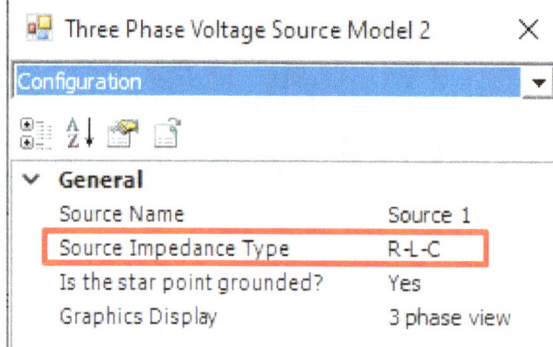

So, head to the configuration menu of R, L and C and configure with the values as below.

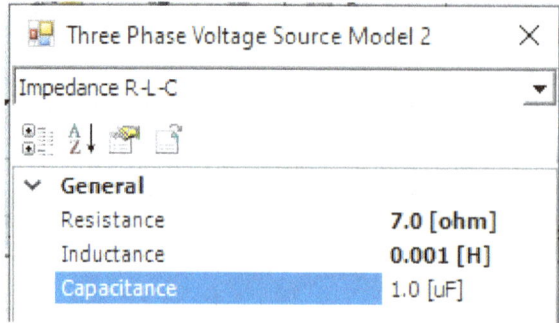

Note:

Each time that you configure parameters, make sure that you also insert the correct unit. This means that in the above configuration user can insert either 0.001 [H] or 1 [mH].

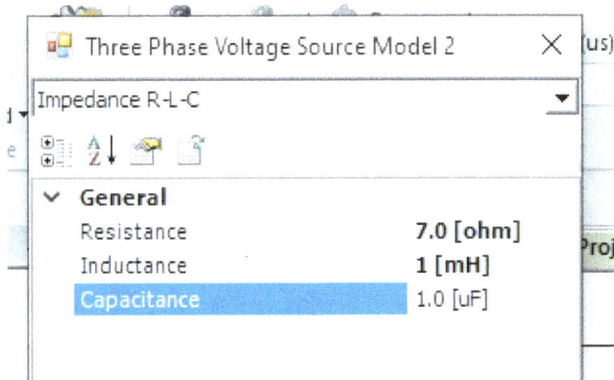

Next, we need to complete the fault. So, to control the fault time we need the so-called timed fault logic. Again, from the Breaker and Fault library select Timed Fault Logic. Then connect to the fault and let's configure like below.

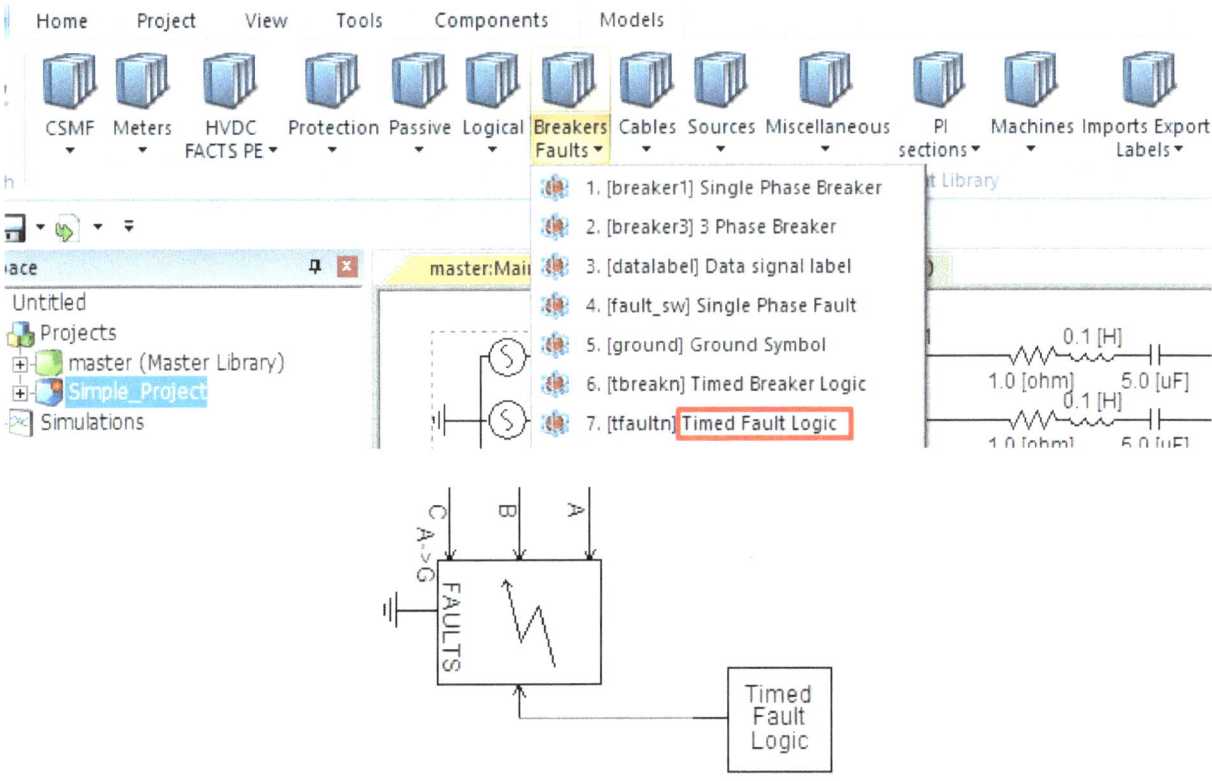

In the configuration menu set the fault time at 3 seconds and the duration of fault 0.2 seconds. To not interfere with the breaker let's set the operation time outside of the duration time of the fault.

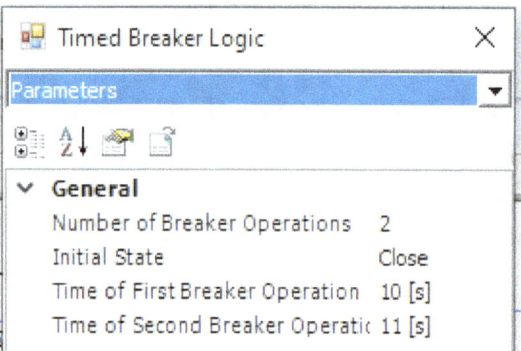

Next configure the fault as below:

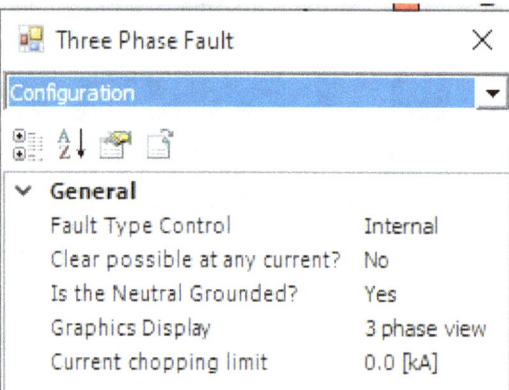

Next configure the phases where the fault occurs. Let's set the fault in the Phases A and B.

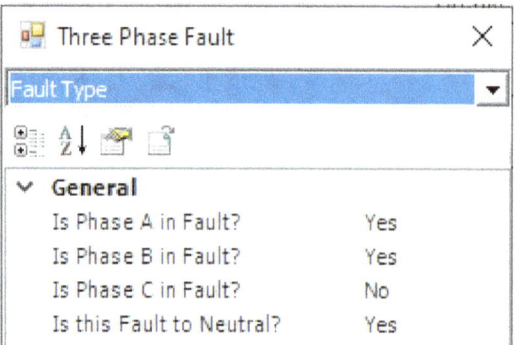

Let's run the simulation and see the results. Make sure to select every plot and click X and then Y to scale the plots automatically.

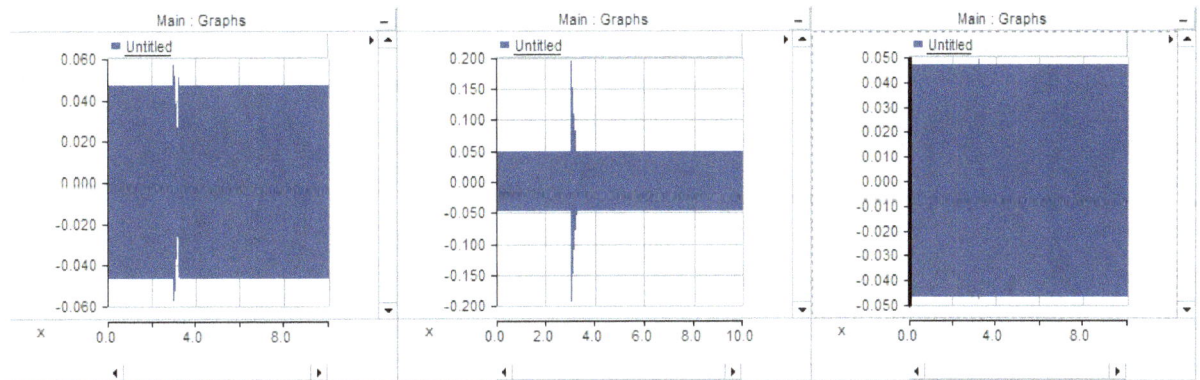

Now re-configure the grid with the ratio of X/R = 7 as below:

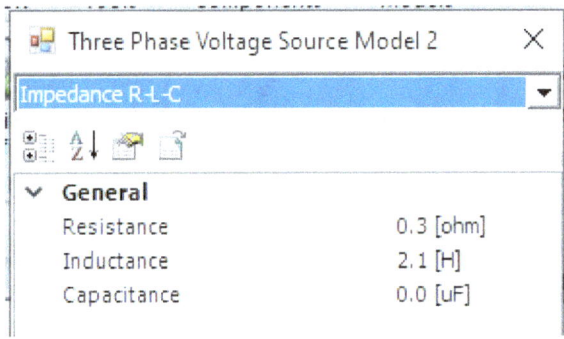

Let's simulate again and view results.

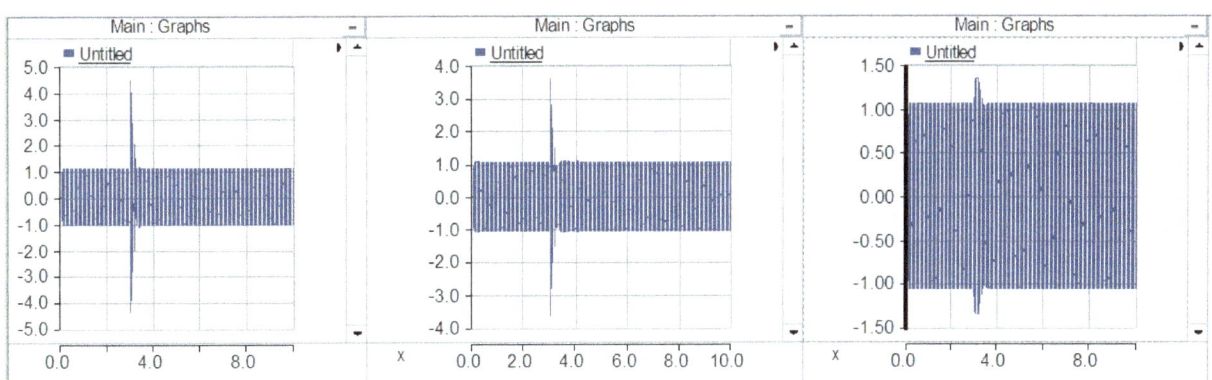

1.9. Sequences

A "Sequence of Events" based on timers, delays, and/or other conditions can be created by combining sequencer components, which are a specific class of control elements. Sequencer components offer an open-ended substitute for the Timed Breaker Logic and Timed Fault Logic control components and are typically utilized as a control mechanism for breakers or faults. Let's see this in more details in our example below.

We will try to simulate a protection scheme for our diagram. The procedure we want to follow is described in the steps below:

1. Start the simulation
2. Wait for a certain amount of time
3. Apply the fault
4. Wait for a certain amount of time
5. Open the breaker

So, we need to perform all the actions controlled and in a sequence order.

- To insert the first step of the sequence head to the Models and then to Sequencer directory and select Sequencer_Start and insert it into the diagram.

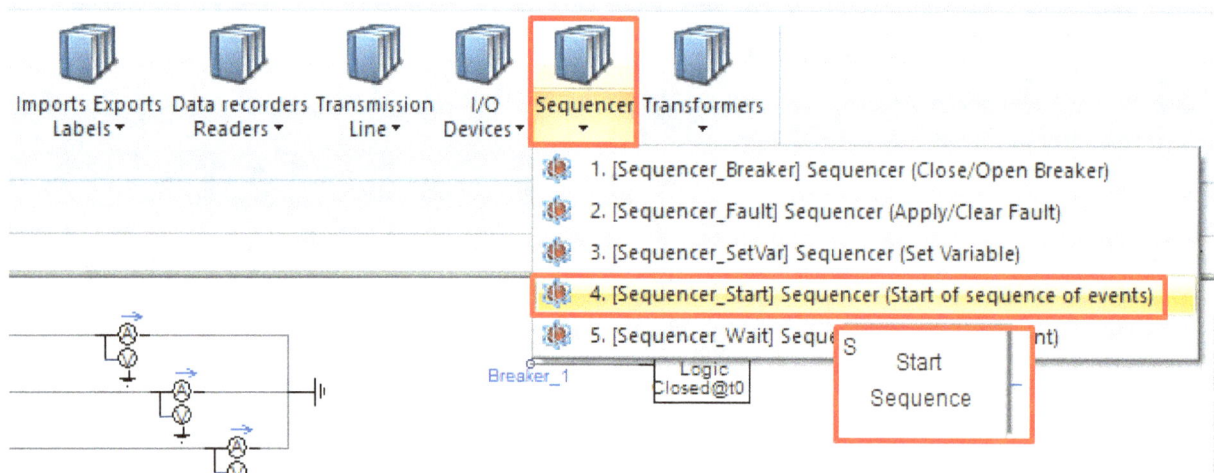

Next double click inserted sequencer and configure the parameters.

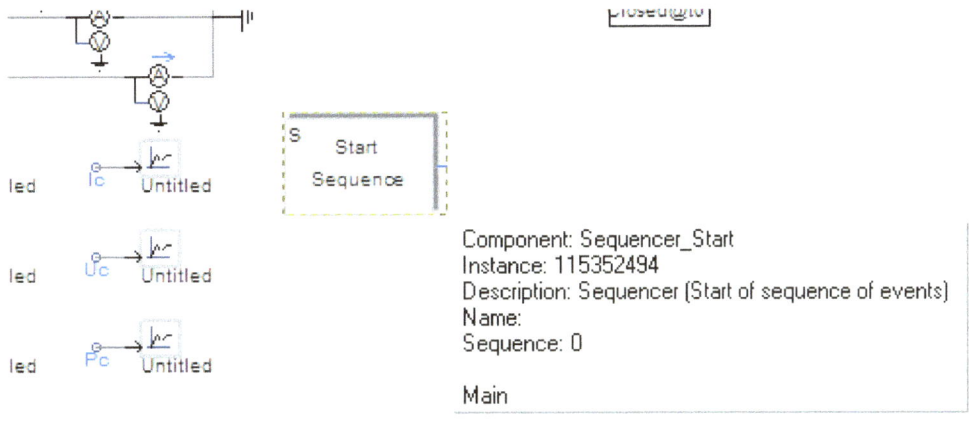

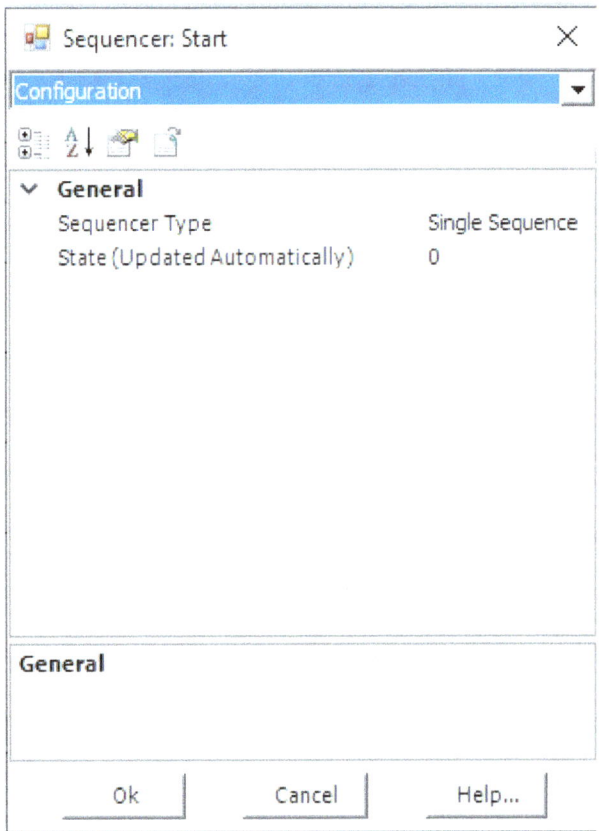

As much as we see, we have only one configuration menu. In the general settings we have 2 settings for the sequence Type. We will go for the default settings which means that we will select "Single Sequence" as we want to be started only once. If we want to put a condition to repeat the sequence we can select "Repetitive Sequence" and set a condition to repeat.

In the next step we want to wait for a certain amount of time before performing the sequence.

Head again to the Models/Sequencer and select "Sequencer_Wait"

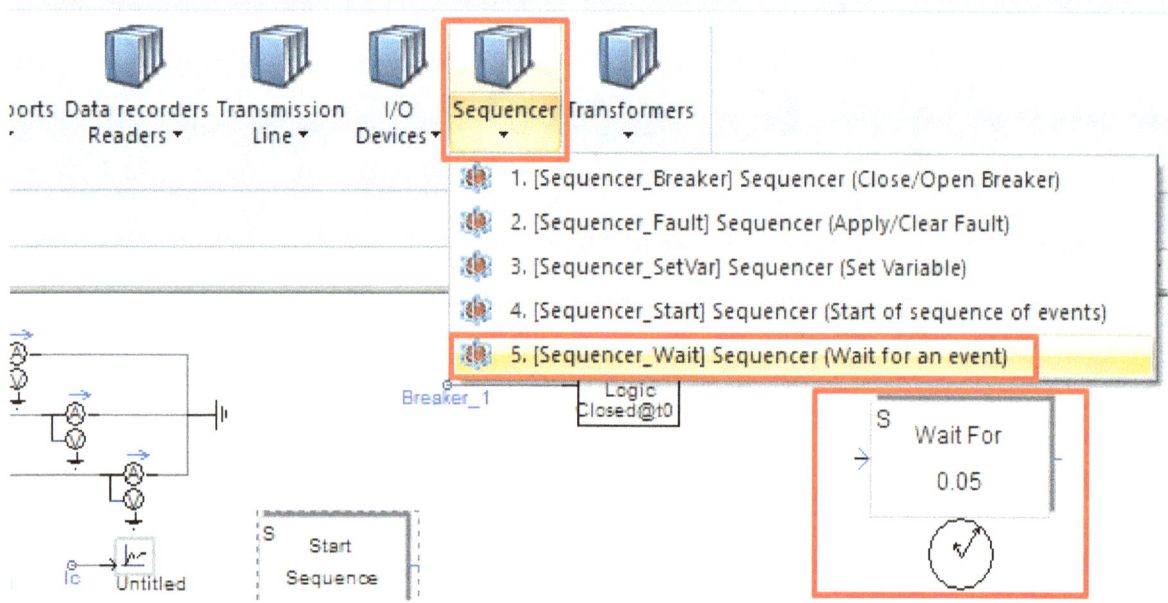

Select the sequencer and connect to the previous sequencer as in the figure below:

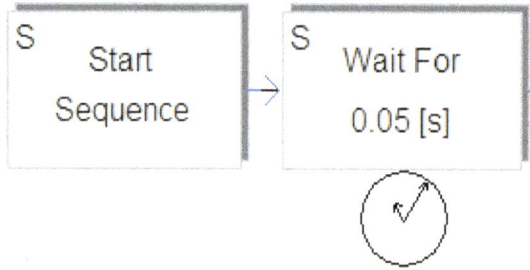

Then configure the sequencer as below:

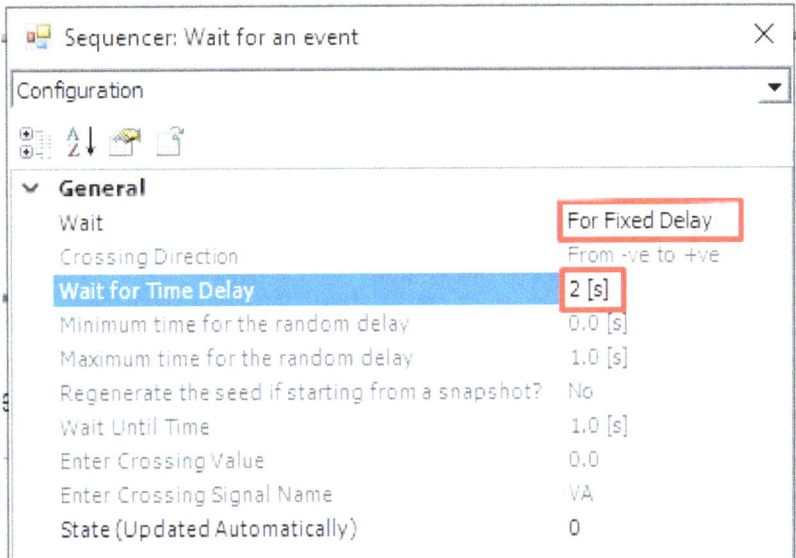

We are going to wait for a Fixed delay. User is free to select between the following configuration of the waiting sequencer:

- Fixed Delay
- Random Delay
- At a specified Unit Time
- For Signal Crossing

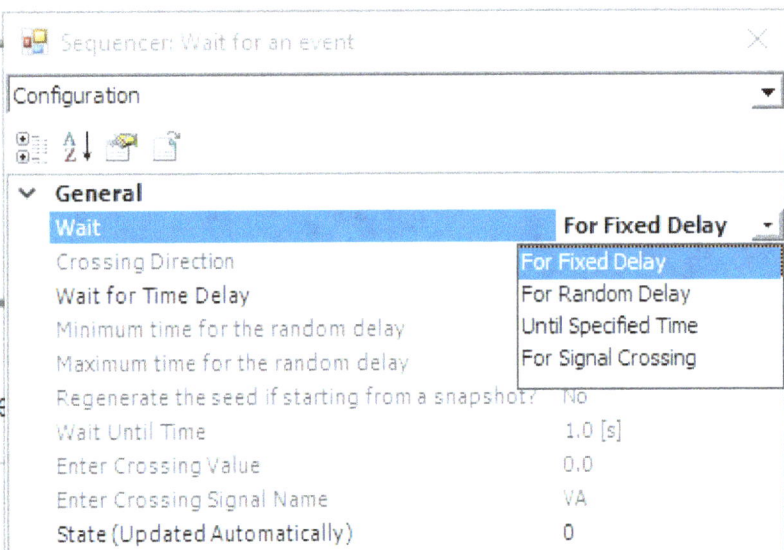

- Now to perform the fault sequence, insert a sequence from Sequencer directory which is called "Sequencer Fault"

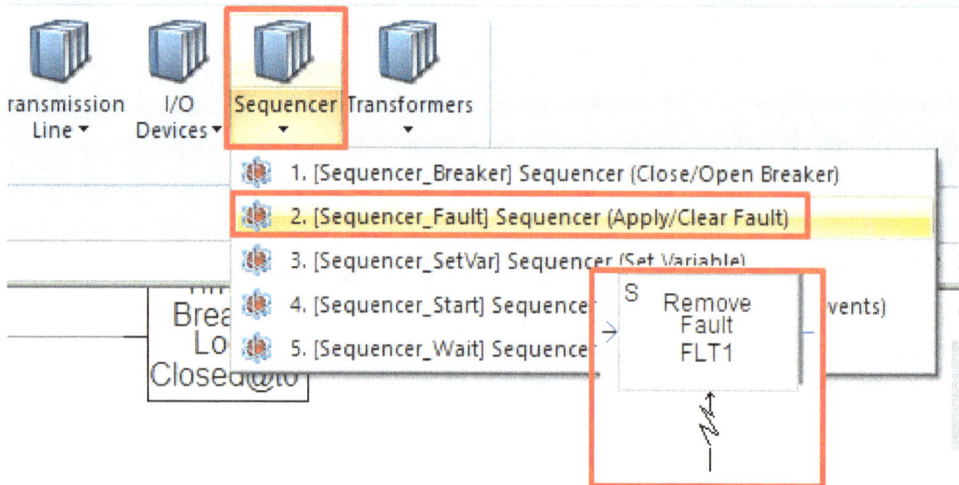

Insert the sequencer next to the previous two, and let's configure the fault sequencer:

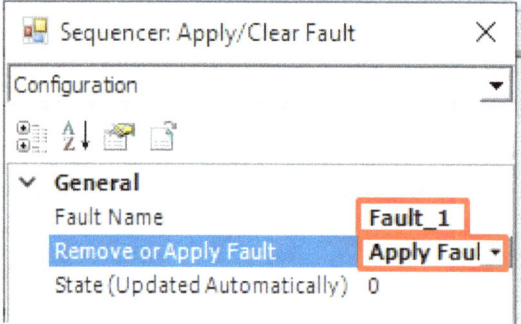

We will apply the fault that we named "Fault_1". The same name that we inserted in the Fault Name, the same name should correspond to the fault block that we want to control. So head to the fault that we have in our diagram and configure the Fault Type Control as "External"

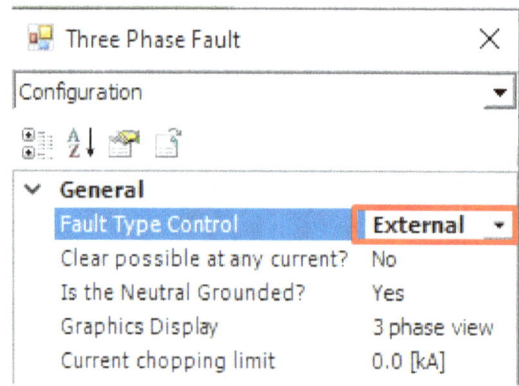

Now insert a label and name it "Fault_1" and connect the external connection created from the Fault block like in the figure below:

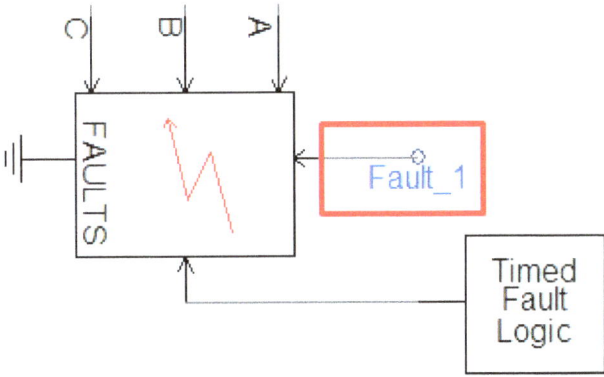

- Next insert a waiting sequence as in the figure below. This would be the time that the breaker should wait before it disconnects the circuit. Insert the sequencer and set it to 1 seconds. Even 1 second is a high risky time in protection field, this is chosen for example purposes. After connecting, the figure should look like below:

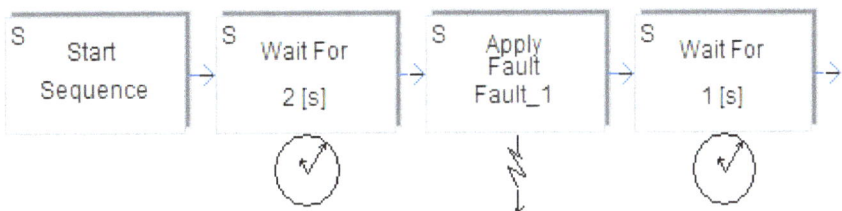

The last step is to open the circuit breaker.

To perform this, we should insert a sequencer that will open the circuit breaker. This can be performed by inserting a "Sequencer_Breaker" as below:

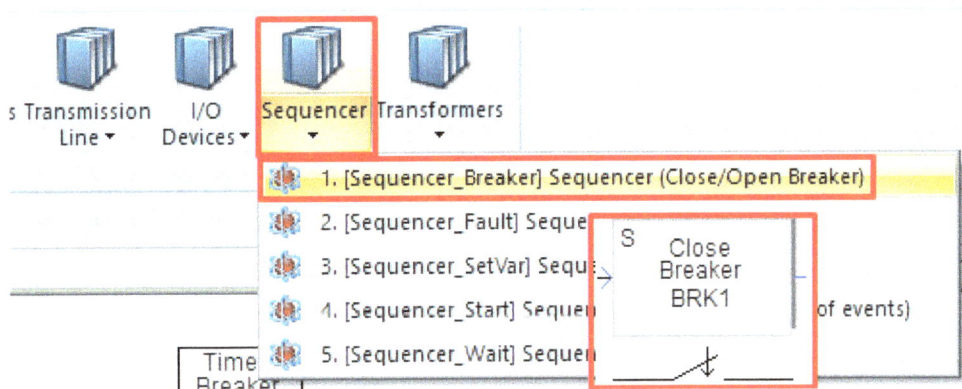

Now let's configure the sequence breaker as below.

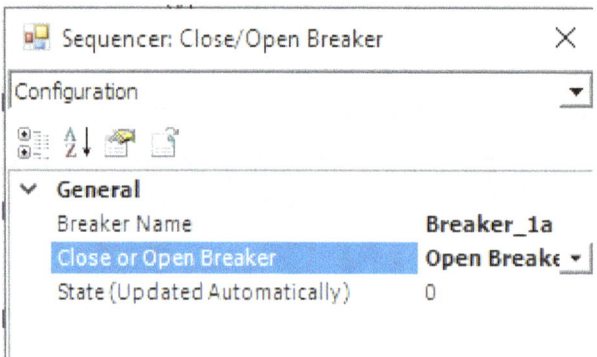

So, we will set the breaker to "Open Breaker" and the breaker name as "Breaker_1a". This is done to avoid conflicts with the breaker name that is set into Timed Breaker Logic.

Then we should rename our breaker name from "Breaker_1" to "Breaker_1a". In this way the breaker behavior will be controlled from our sequence breaker and not from our Timed Breaker Logic. Also, user can set the same name and delete the Timed Breaker Logic block.

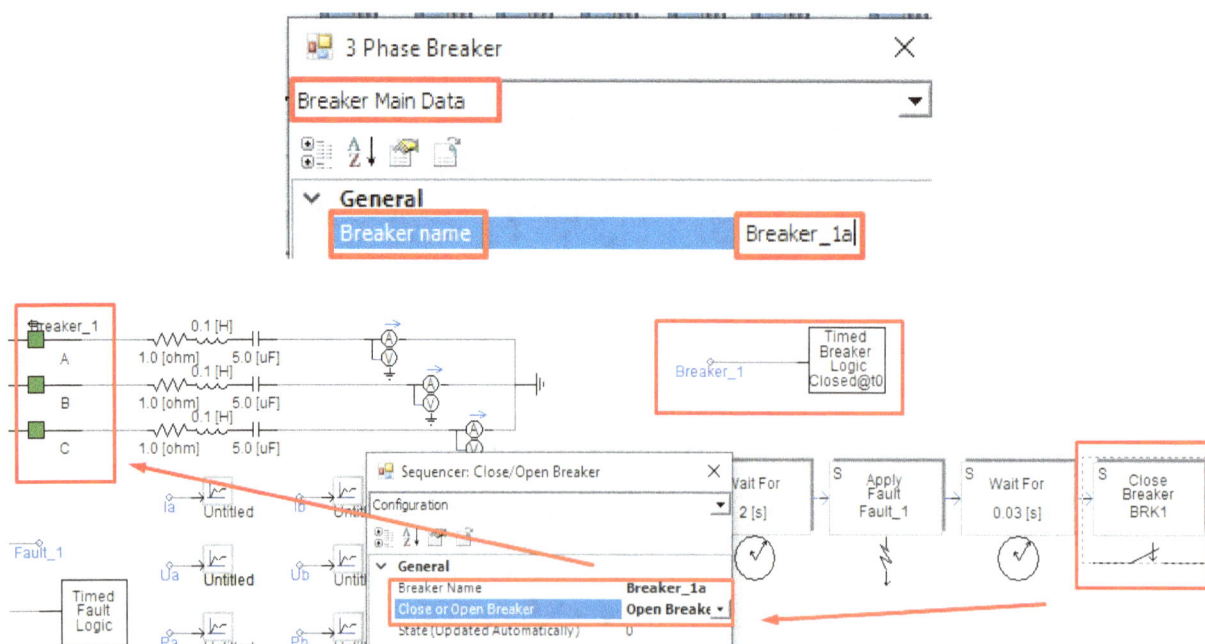

Now it is time to run simulation. Before running we need to change the time of simulation in order to view the results much clearer. So, the simulation time for our example will be 5 seconds. Let's simulate and analyze the results.

Zoom with the rectangle zoom in the plots below:

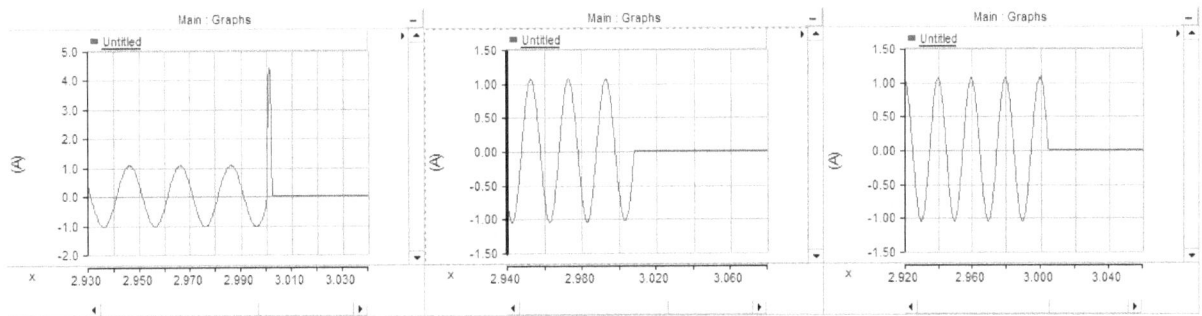

So, we see that after 2 seconds we have the fault and the disconnection of the breaker after 2+1 seconds.

1.10. Modeling Wind Turbine Generator

Now after learning the most fundamentals and being confident on PSCAD environment, it is time to jump to a more advanced topic regarding power system. We will create a wind turbine connected to a permanent magnet synchronous generator and then connected to the grid via a fully bridged converter AC/DC/AC. Also in this section, beside simulating the wind turbine, we will go through DC elements, synchronous machines and some other features of PSCAD.

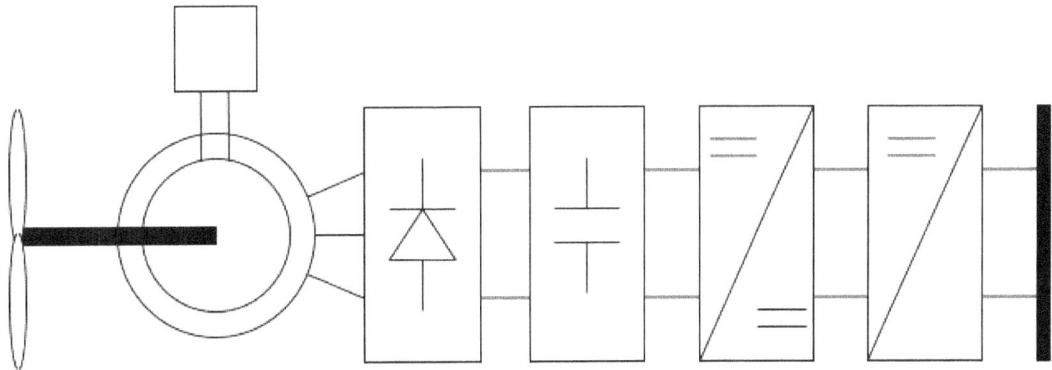

Before continuing building the model, make sure that the setting of enabling users to use different type of units beside those default ones.

This can be done by heading to project settings by right clicking in the diagram that we are working and then in the Dynamics tab and then check "Enable unit conversion and apply to parameter values". This will allow users to use different units during modeling.

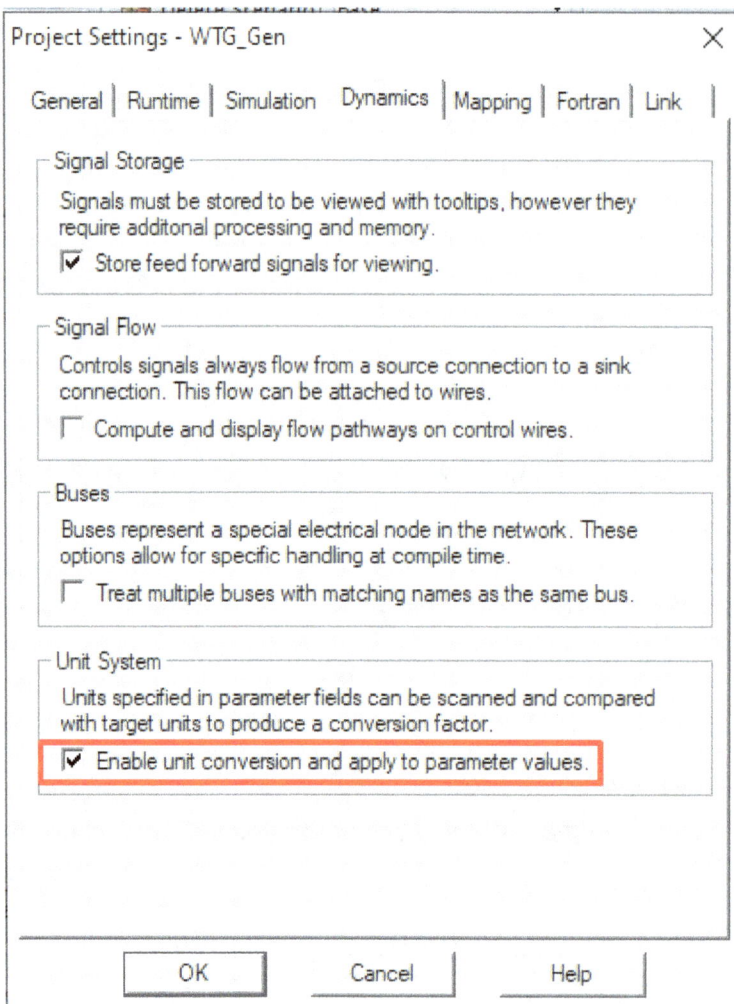

1.11. Wind turbine components

Below we will list all the components that are required to build a simple wind turbine. In PSCAD a wind turbine is compounded by three main parts:

- Wind Source
- Wind Turbine component
- Wind turbine governor

1.11.1. Wind Source

Wind source can be found under Machines/wind source as in the figure below

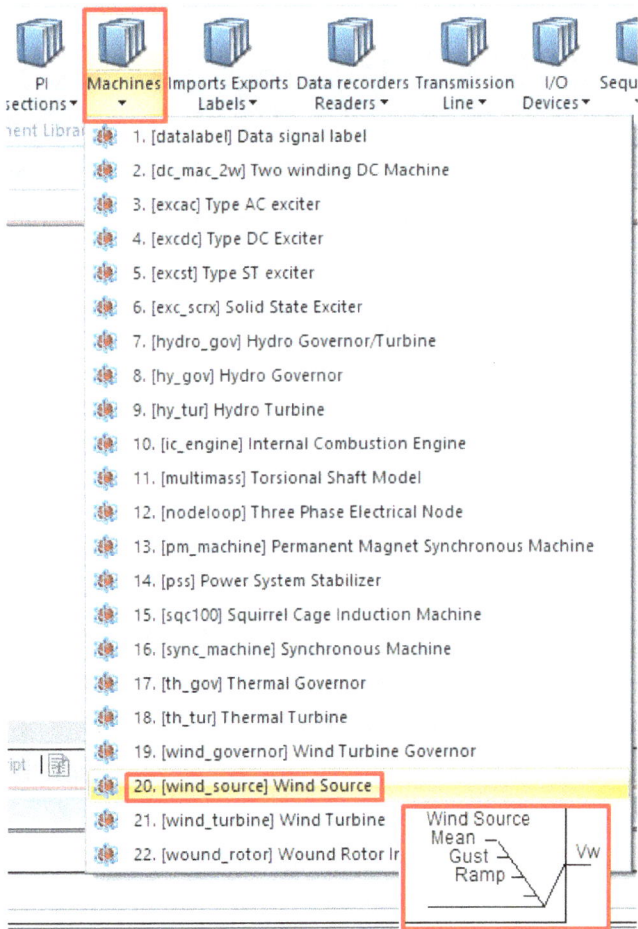

Let's analyze the block configuration data as below:

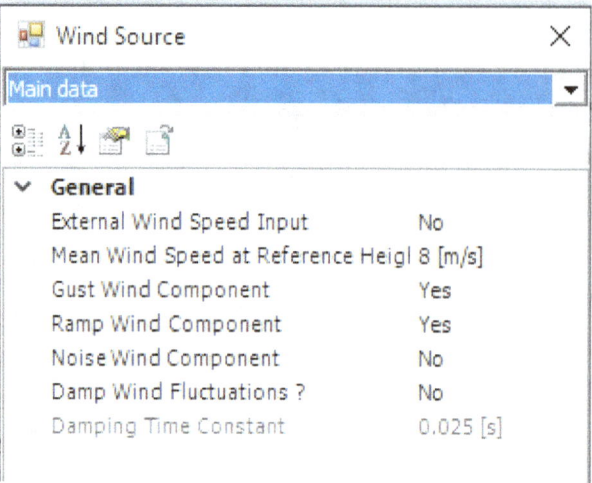

The wind data block contains all relevant information that is needed for modeling a wind. The data can be inserted from external measurements in form of text files or by manually inputting them in the block above. So, we will go for static wind speed of 13m/s as this is the nominal wind speed needed to produce nominal power from a wind turbine. Normally a wind turbine would start at 5m/s and it will reduce speed at a cut out speed of 25 m/s, called also as the critical speed risking wind turbine to lose synchronism. We will use a pure linear wind speed so no Gust, ramp or noise configuration here.

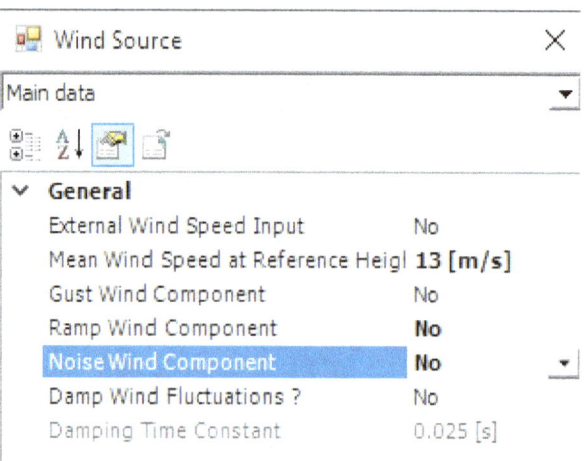

1.11.2. Wind turbine component

Wind turbine can be inserted from Machines/wind turbine.

This component simulates the basic mechanics of a wind turbine by taking into account the number of blades (2 or 3), tip speed ratio, coefficient of power, and sweep area and radius. This model, which may be used in conjunction with the wind governor model, does not take into account shaft dynamics. The inputs consist of the mechanical speed of the machine attached to the turbine (W) and the wind speed (Vw). The turbine blades' pitch angle, or beta, is specified in degrees. Based on the machine rating, Tm and P represent the output torque and power, respectively, in per-unit terms. Connect both blocks as in the figure below:

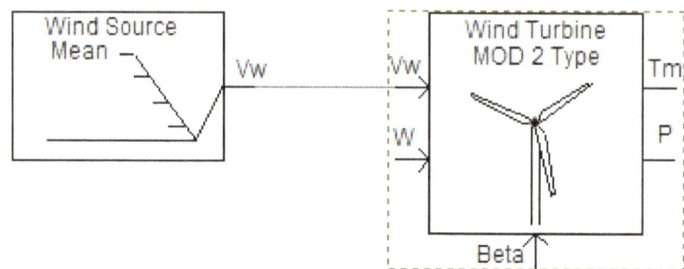

Now let's model the block inside. First, we will need some theorical explanations of what are we going to model and how did we extracted the parameters that we will insert to the block.

Through the rotor blades the kinetic energy that flows is:

$$E = \frac{1}{2} * m * W * s^2$$

Then we can extract a power from wind turbine that is theoretical:

$$P = \frac{1}{2} * \rho * S * s^3$$

Where:

$$\rho = air\ density\ \left(1.22k\frac{g}{m^3}\right)$$

$$S = rotor\ surface\ m^2$$

$$W = wind\ speed\ m/s$$

In more real terms due to the losses of the air that we have because we know that the air that hits the blades doesn't have the speed = 0 as it is after the blades. For this reason, not all the power is extracted from the air. For that we have a coefficient called Betz (from Bernouilli equation) β

This coefficient is calculated as below:

$$\beta = \frac{P_{real}}{P}$$

$$\beta = \frac{1}{2} * (1 - a^2)(1 + a)$$

In our example we will use a rotor with 3 blades and so the coefficient is smaller than its theoretical value. The user can check the formulas by right clicking Help in any block and see the details inside. The power coefficient is described as:

$$\beta = 0.5 * (\lambda - 0.022 * a^2 - 5.6) * e^{-0.17\lambda}$$

Where:

- $\lambda = 2.237 * \dfrac{Wind\ speed}{hub\ speed}$

- $\alpha = incidence\ angle\ of\ the\ blade$

Now let's calculate the parameters for the block. We will use a permanent magnet synchronous machine.

- Number of pole pairs   p = 120
- Rated speed at 50 Hz   n = 2*π*f/p = 2*3.14*50/120 = 2.61 rad/s
- Rated Power Sn = 2 MVA
- Rated voltage = 0.4 kV
- Xd = 0.5
- Rated current In = Sn/(3*Un) = 2000000/ (3*400) = 1666.6 A

- Let's calculate the β as we will need to calculate the radius of the turbine and the rotor area.

At the rated conditions we will connect directly to the turbine without a gearbox (for the sake of the simplicity of the model) $\alpha^2 = 0$ and the hub speed is the same with the synchronous generator rated speed. The wind speed needed to achieve the nominal power is 13m/s.

$$\lambda = 2.237 * \frac{Wind\ speed}{hub\ speed} = 13 * \frac{2.237}{2.61} = \mathbf{11.142}$$

$$\beta = 0.5 * (11.142 - 0.022 * 0^2 - 5.6) * e^{-0.17*11.142}$$

$$\boldsymbol{\beta = 0.4}$$

- And the turbine rated power is calculated like below:

Looking forward the fact that turbine is connected directly to the rotor without a gearbox and therefore we have a friction in the mechanical cycle, the turbine rated power is 20% more than the generator:

Snturbine = Sn * 1.2 = 2*1.2 = 2.4 MVA

- Rotor radius and area:

In PSCAD the power formula for the wind turbine is

Pt = 0.5 * ρ * S * W^3 * β = 0.5 * 1.22 * S * 13 * 0.4

S $\;=\;$ 2400000 / (0.5 * 1.22 * 13^3 * 0.4) = 5462m2

And the radius by the equation of $\mathbf{S = \pi r^2}$ ➔ 41.7m

1.11.3. Wind Turbine Governor

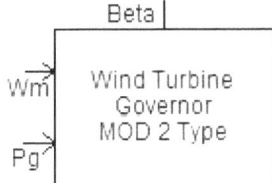

This part simulates a wind turbine's pitch angle regulator. The mechanical speed of the machine (Wm) and the power output (Pg) of the machine are the model's inputs. The turbine's pitch angle is the result.

From the power coefficient formula from above:

The regulation of

$$\beta = 0.5 * (\lambda - 0.022 * \alpha^2 - 5.6) * e^{-0.17\lambda}$$

The regulation of α^2 enables the regulation of β and in this way, we can control the output power of the wind turbine. This is also dependent on the wind power. Here we can have 2 strategies that may apply:

- Passive pitch control

The wind turbine manufacturer chooses the angle α^2 to produce the most energy at a predetermined average speed. There is no angle control below the mean wind speed. Beta isn't at its highest. Turbulence is produced by the blade profile above the mean wind speed to prevent the blades' rotation from accelerating.

- Dynamic pitch control

Dynamic Pitch Control: The blades in this arrangement may rotate about their longitudinal axes. A power reference is provided for the regulation system, and it spins the blades once every second to control the output power as seen on the following curve.

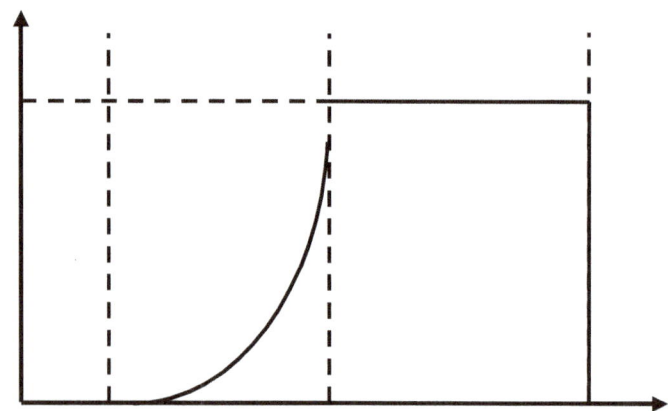

1.11.4. Governor properties PSCAD

Now let's define the governor properties as below:

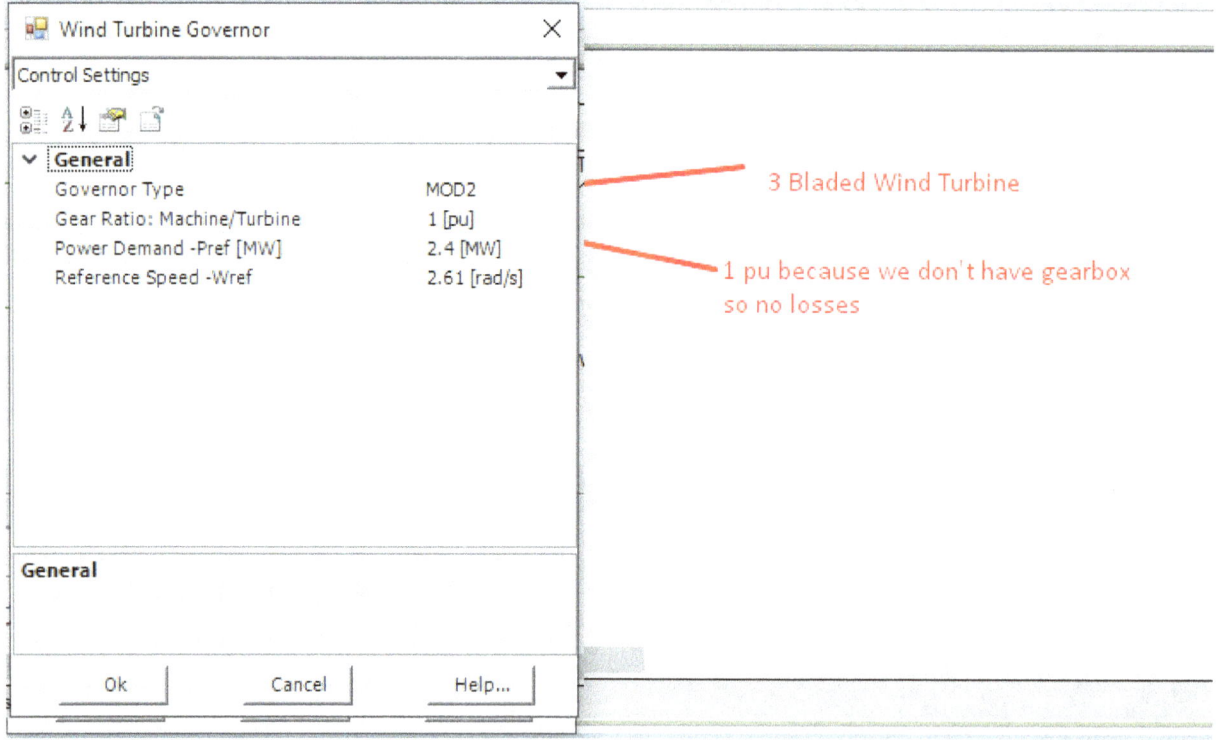

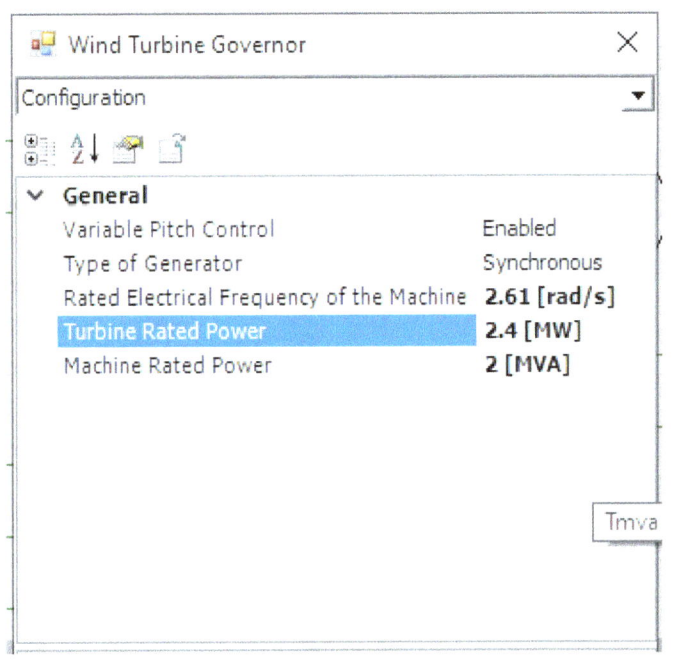

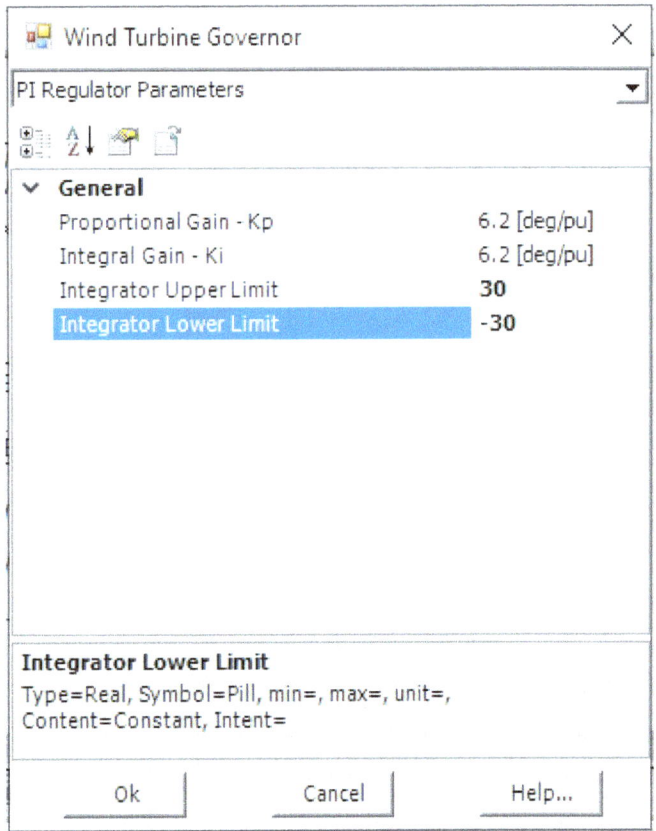

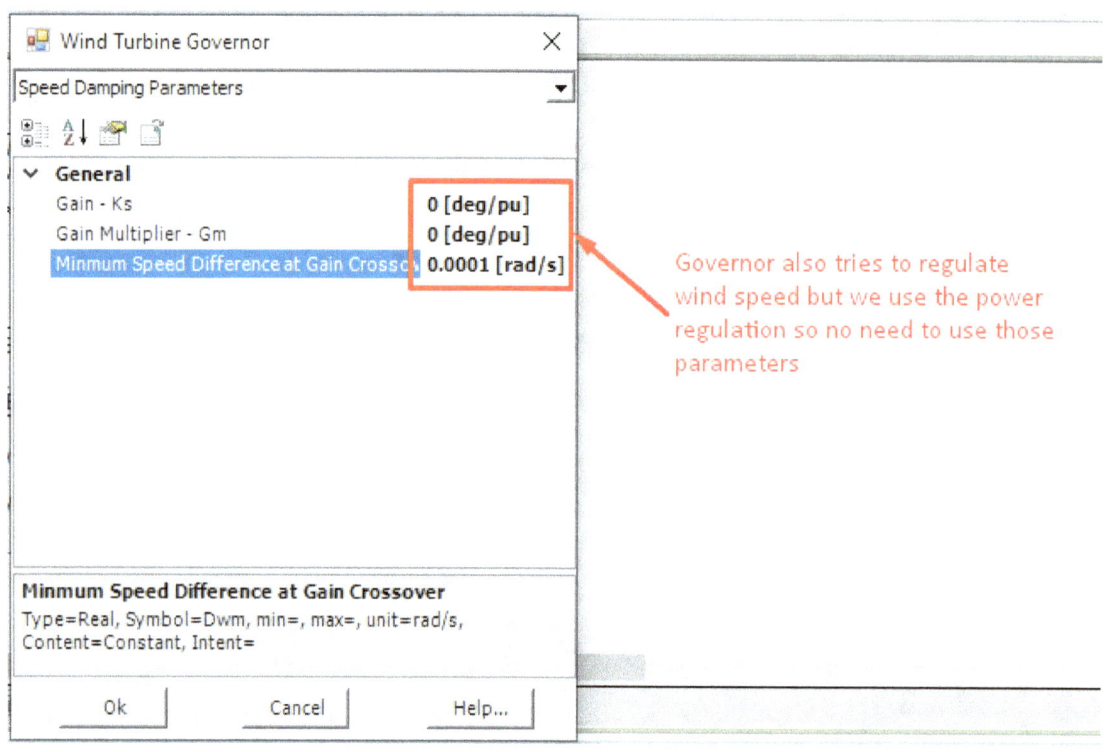

Governor also tries to regulate wind speed but we use the power regulation so no need to use those parameters

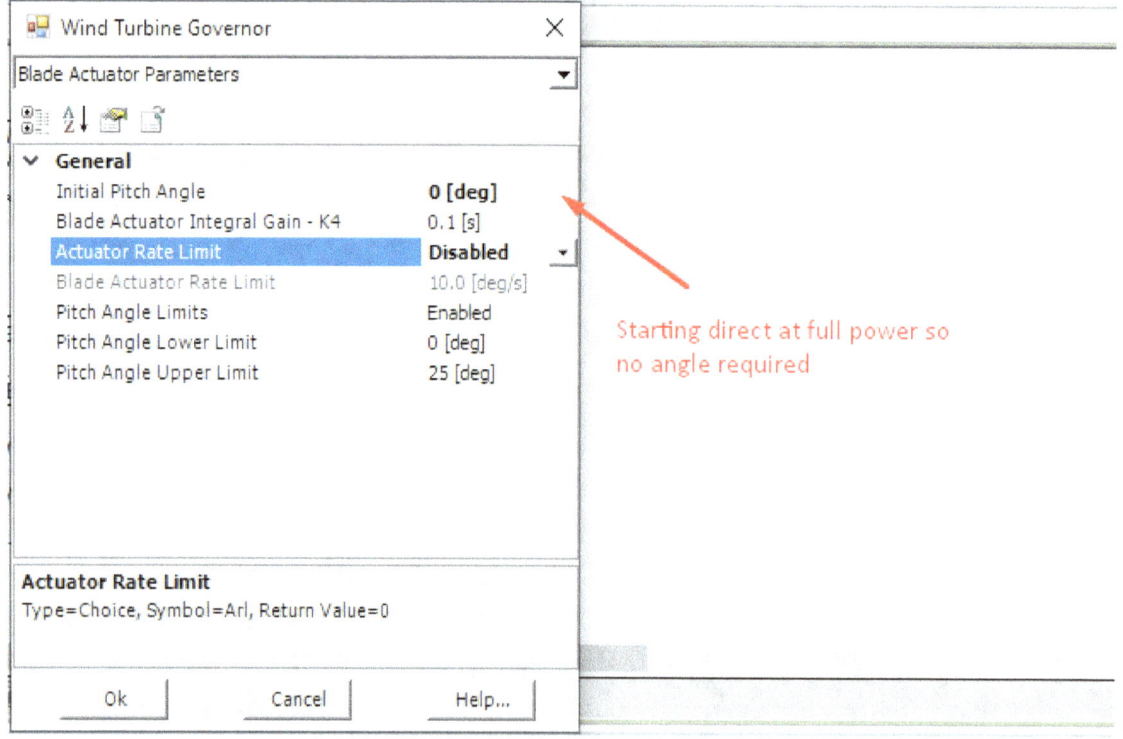

Starting direct at full power so no angle required

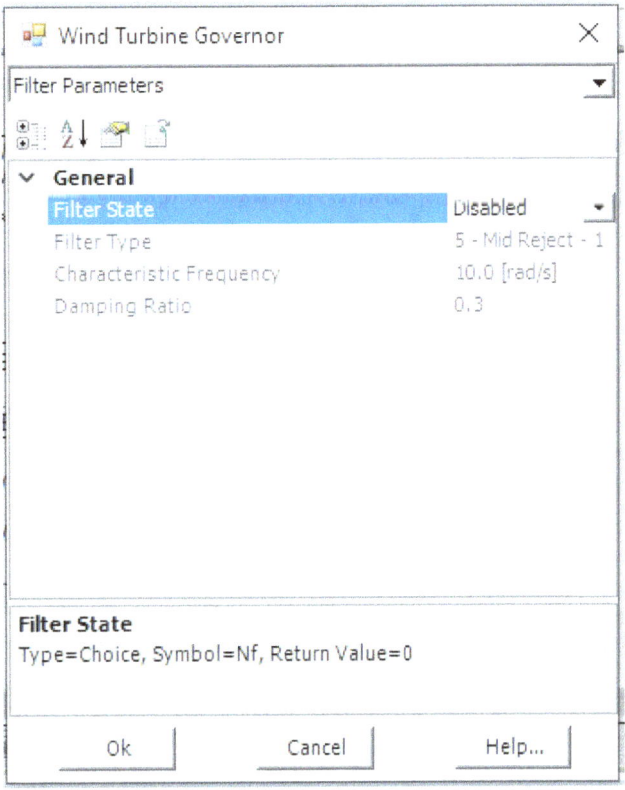

1.11.5. Synchronous Generator

This part may be used as either a salient pole machine or a round rotor machine since it has the option to simulate two damper windings in the Q-axis. By entering a positive number into the machine's w input or by applying a mechanical torque to the Tm input, the speed of the machine may be directly controlled.

This component has numerous sophisticated features for simulating synchronous machines. For typical use, the "Advanced" parameters can be kept at their default settings without affecting the machine's anticipated performance. These characteristics primarily seek to fast obtain the necessary steady state and initialize the simulation.

Machine parameter descriptions are more commonly used in this input data entry style. Here, per-unit values are entered for variables like Xd, Xd', Xd", Tdo', Tdo", etc.

Some producers offer armature resistance while others offer the time constant. You can enter the armature resistance in per-units or the time constant in seconds if the Data Entry Format input option is set to Generator.

In our model of our synchronous generator the following notes apply:

- The excitation voltage we have chosen is stable – 1pu
- Large unsaturated transient time Tdo' which increases the field leakage: 10 s
- Very small unsaturated time transient Tdo" and thus we have a high dumper resistance: 0.0001

The synchronous machine can be inserted from machines like below:

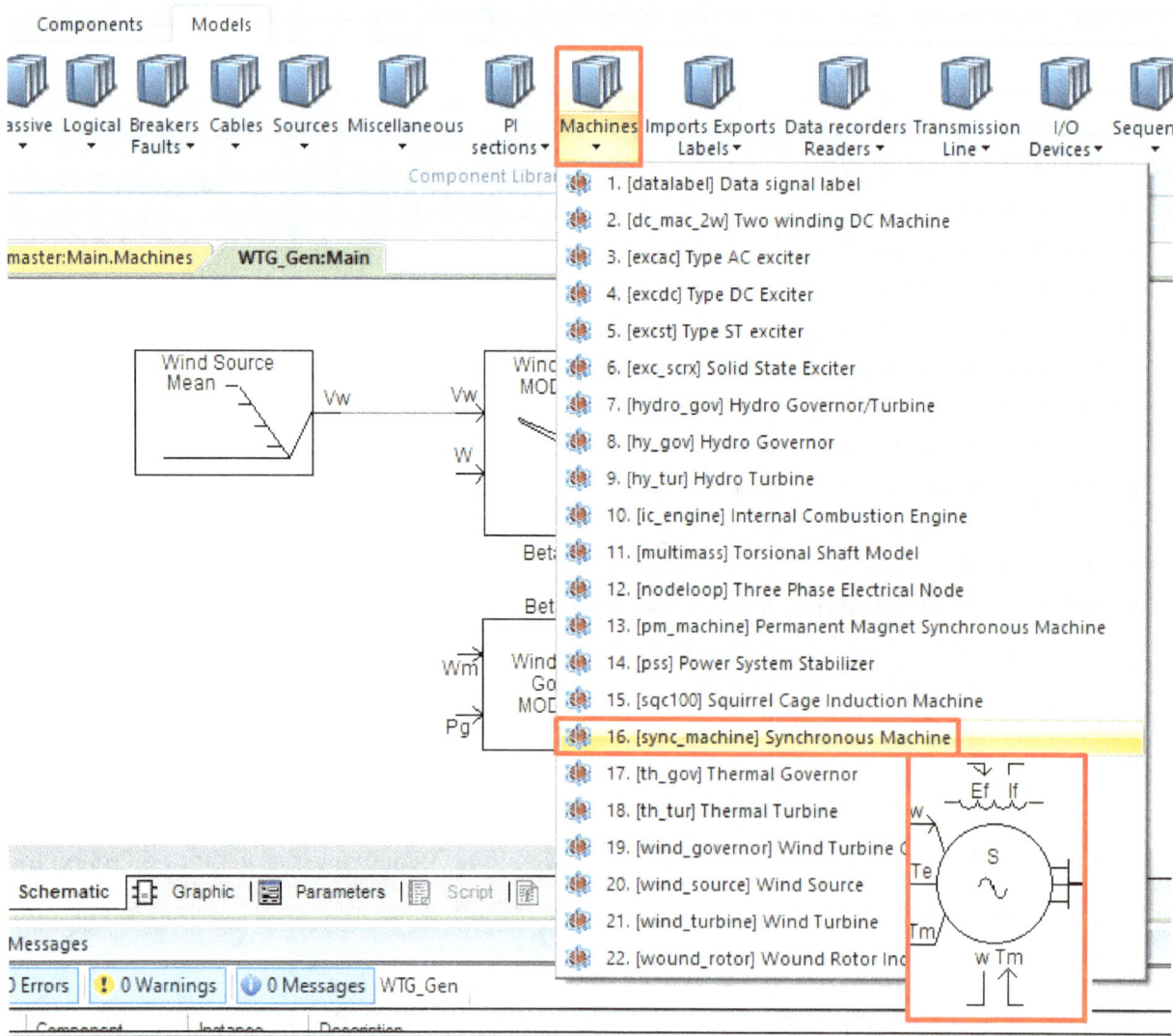

Now knowing the upper explanations let's try to model the synchronous machine:

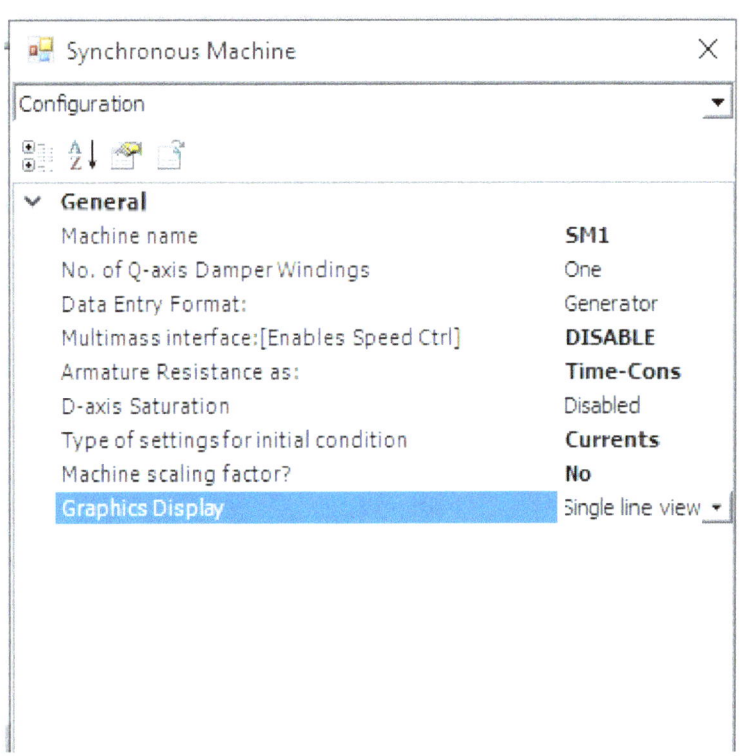

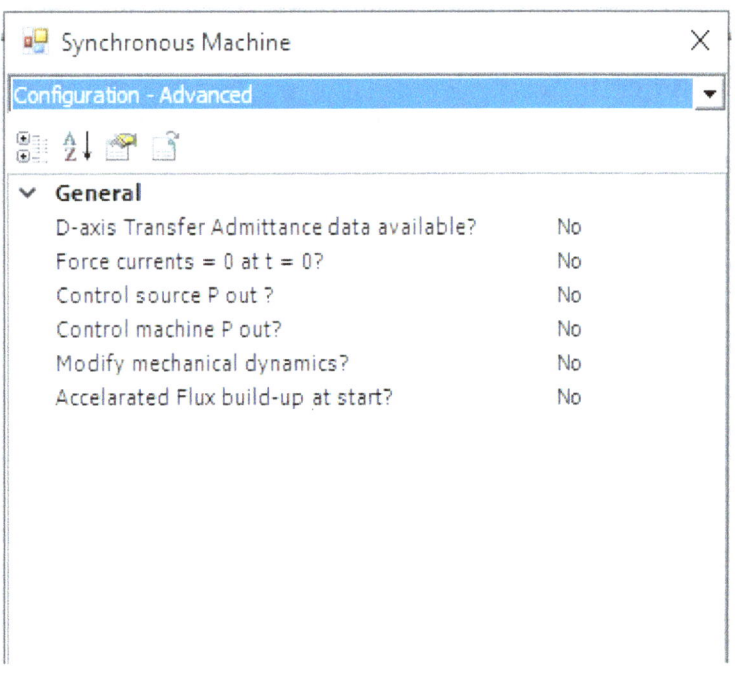

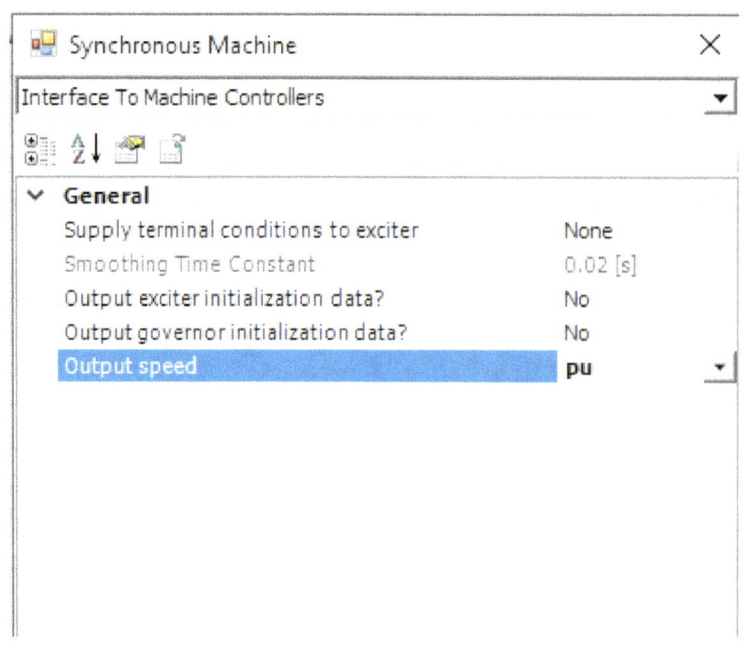

Synchronous Machine ✕

Interface To Machine Controllers ▼

General
Supply terminal conditions to exciter	None
Smoothing Time Constant	0.02 [s]
Output exciter initialization data?	No
Output governor initialization data?	No
Output speed	**pu**

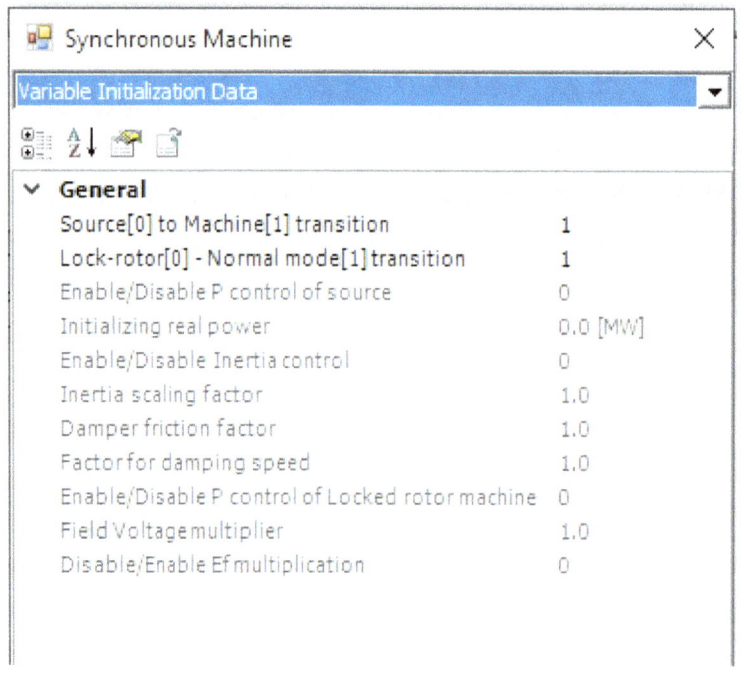

Synchronous Machine ✕

Variable Initialization Data ▼

General
Source[0] to Machine[1] transition	1
Lock-rotor[0] - Normal mode[1] transition	1
Enable/Disable P control of source	0
Initializing real power	0.0 [MW]
Enable/Disable Inertia control	0
Inertia scaling factor	1.0
Damper friction factor	1.0
Factor for damping speed	1.0
Enable/Disable P control of Locked rotor machine	0
Field Voltage multiplier	1.0
Disable/Enable Ef multiplication	0

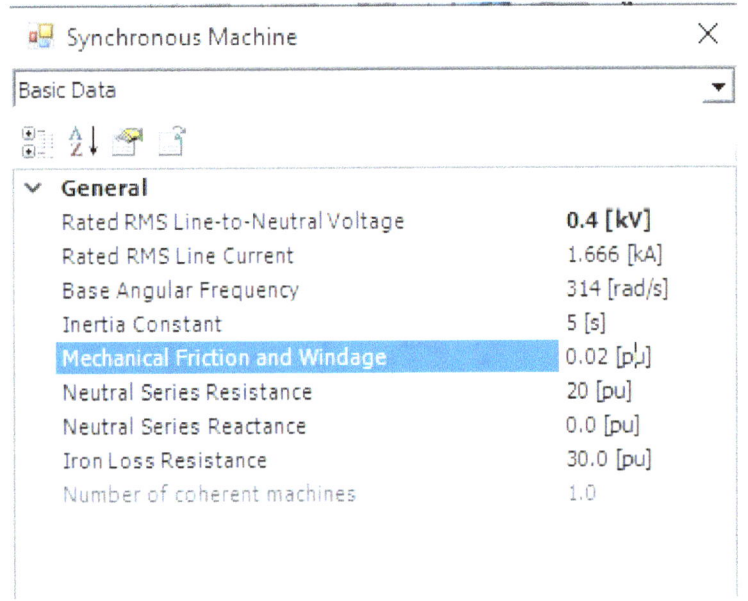

Synchronous Machine ✕

Basic Data ▾

General
Rated RMS Line-to-Neutral Voltage	**0.4 [kV]**
Rated RMS Line Current	1.666 [kA]
Base Angular Frequency	314 [rad/s]
Inertia Constant	5 [s]
Mechanical Friction and Windage	0.02 [pu]
Neutral Series Resistance	20 [pu]
Neutral Series Reactance	0.0 [pu]
Iron Loss Resistance	30.0 [pu]
Number of coherent machines	1.0

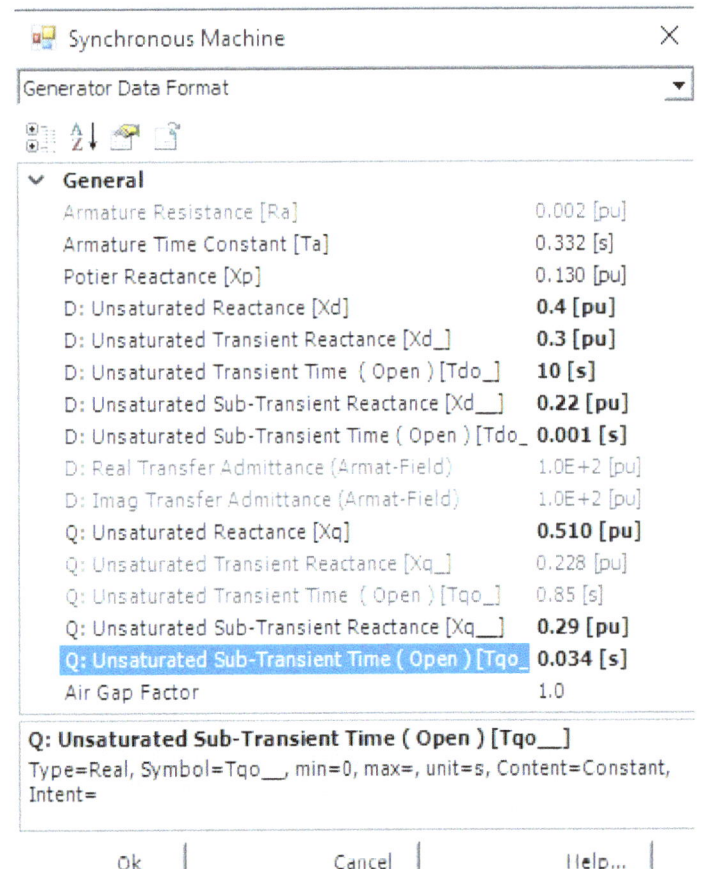

Synchronous Machine ✕

Generator Data Format ▾

General
Armature Resistance [Ra]	0.002 [pu]
Armature Time Constant [Ta]	0.332 [s]
Potier Reactance [Xp]	0.130 [pu]
D: Unsaturated Reactance [Xd]	**0.4 [pu]**
D: Unsaturated Transient Reactance [Xd_]	**0.3 [pu]**
D: Unsaturated Transient Time (Open) [Tdo_]	**10 [s]**
D: Unsaturated Sub-Transient Reactance [Xd__]	**0.22 [pu]**
D: Unsaturated Sub-Transient Time (Open) [Tdo_	**0.001 [s]**
D: Real Transfer Admittance (Armat-Field)	1.0E+2 [pu]
D: Imag Transfer Admittance (Armat-Field)	1.0E+2 [pu]
Q: Unsaturated Reactance [Xq]	**0.510 [pu]**
Q: Unsaturated Transient Reactance [Xq_]	0.228 [pu]
Q: Unsaturated Transient Time (Open) [Tqo_]	0.85 [s]
Q: Unsaturated Sub-Transient Reactance [Xq__]	**0.29 [pu]**
Q: Unsaturated Sub-Transient Time (Open) [Tqo_	**0.034 [s]**
Air Gap Factor	1.0

Q: Unsaturated Sub-Transient Time (Open) [Tqo__]
Type=Real, Symbol=Tqo__, min=0, max=, unit=s, Content=Constant, Intent=

| Ok | Cancel | Help... |

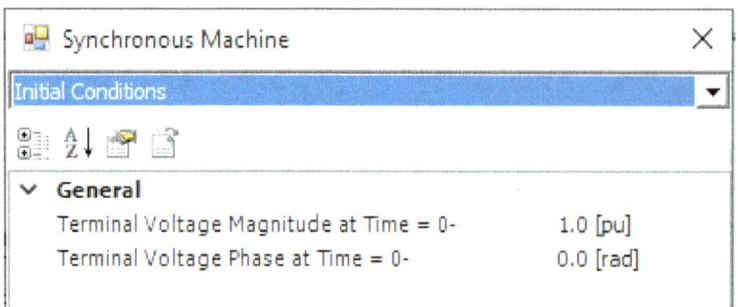

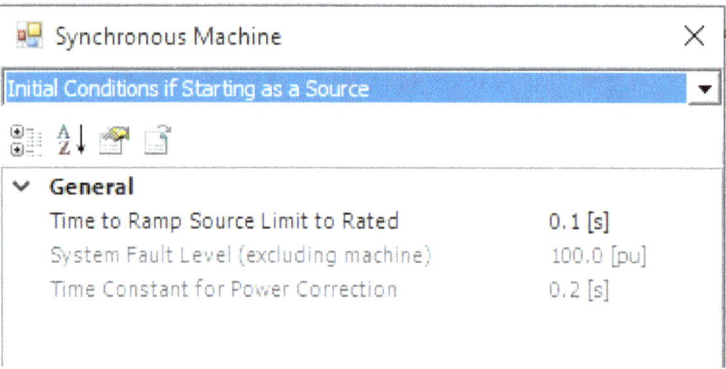

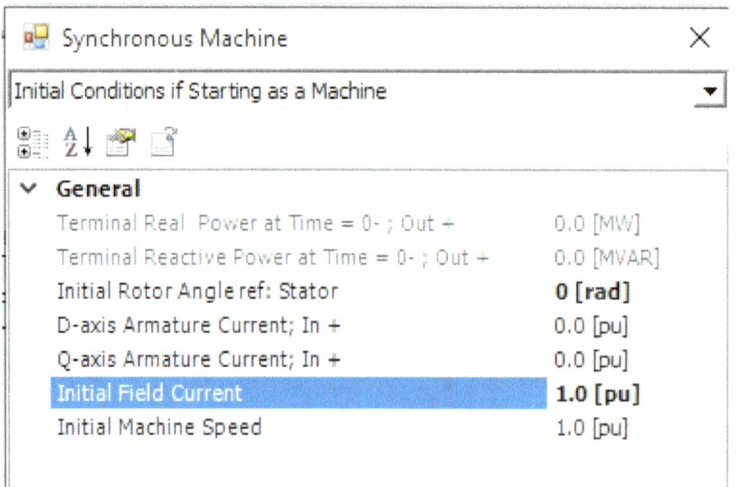

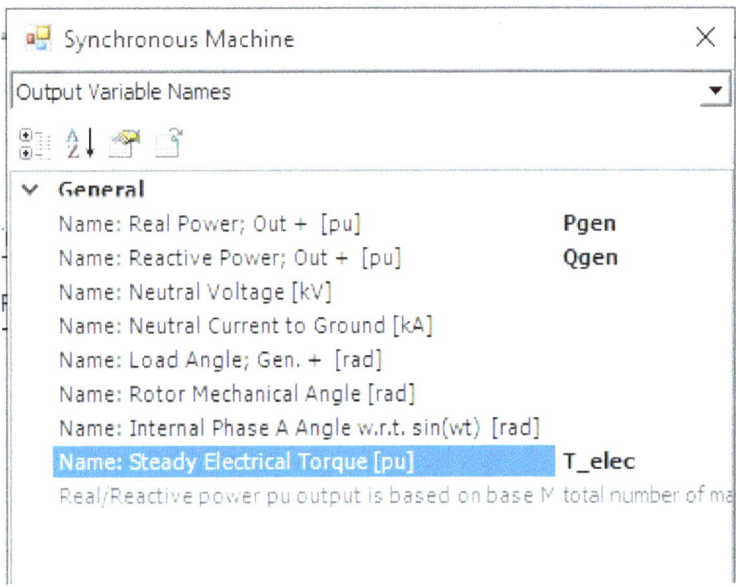

Now let's connect all the elements that we inserted in order to obtain a full wind turbine component like in the figure below:

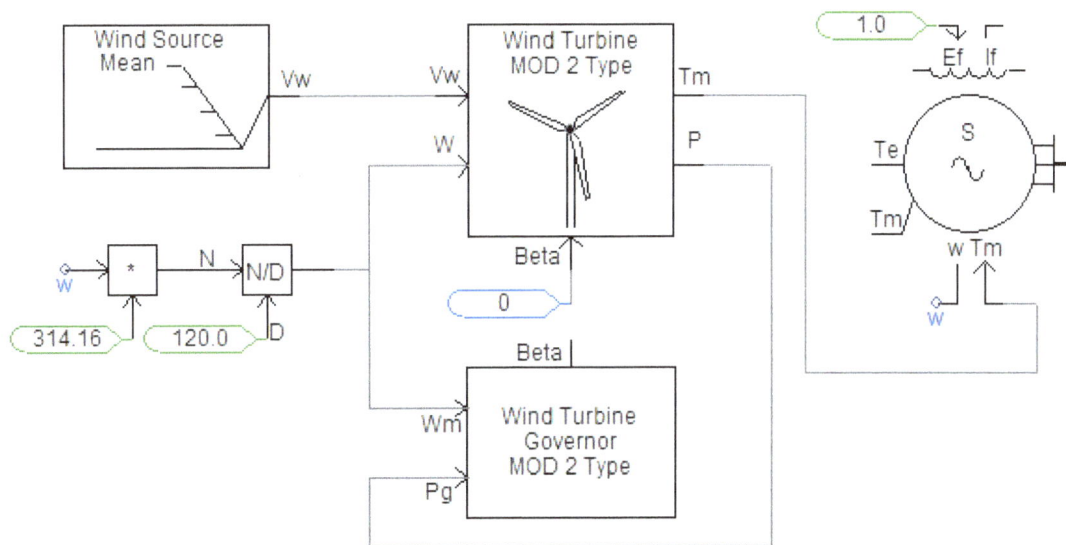

In the figure above we have this block like in the figure below which will create the mechanical speed for the turbine and the governor. This is equivalent to 2*п *f / n = 2*3.14*50 / 120

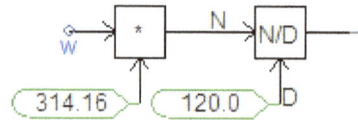

Wind source is connected to the turbine model by providing the wind speed and the initial angle of course will be 0.

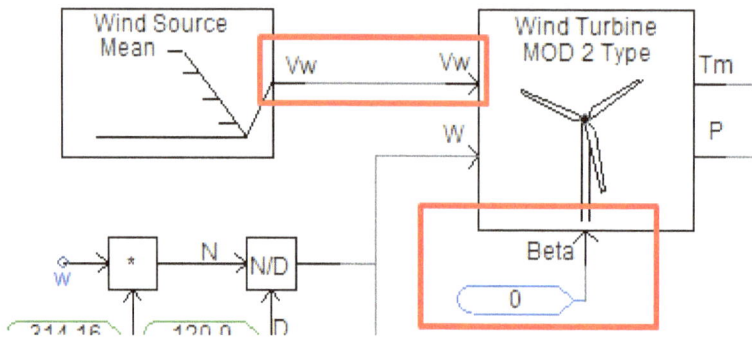

The wind turbine will produce a mechanical torque, which will be passed as an input to our synchronous machine. The synchronous machine will start with a static excitation field defined by the constant of value 1 inserted at the port Ef.

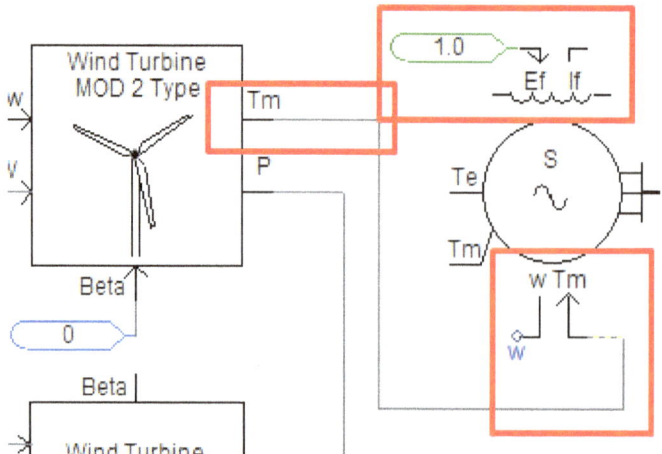

Also, the mechanical speed is connected to the synchronous machine.

Now it is time to connect the wind turbine to a rated load. The load will be represented by a resistance R which can be calculated as below:

$$P = 3 * U * I * cos\emptyset = 3 * U * \frac{U}{Z} * \frac{R}{Z} = \frac{3 * V^2 * R}{(R^2 + Xd^2)}$$

From above equations we have Xd = 0.5

From the equation below:

$$P = \frac{3 * V^2 * R}{(R^2 + Xd^2)}$$

Or by simply calculating Xd =0.5pu = U/I = 0.5 * 480/1666.6 = 0.14 Ohm

Then select a resistance and connect like in the figure below:

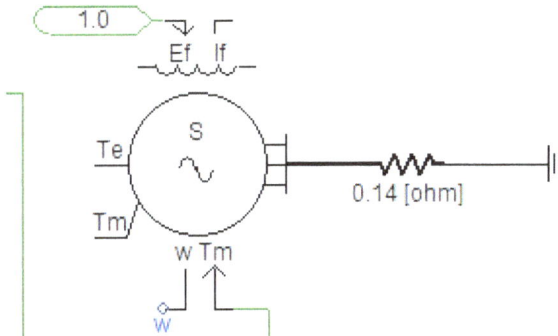

Now let's try to run a simulation. We will need some variables to observe which are:

- Active Power Generation (provided from machine)
- Reactive Power Generation (provided from machine)
- Electromagnetic Torque (provided from machine)
- Mechanical Speed
- Turbine Torque

The variables observed should be taken like in the figure below:

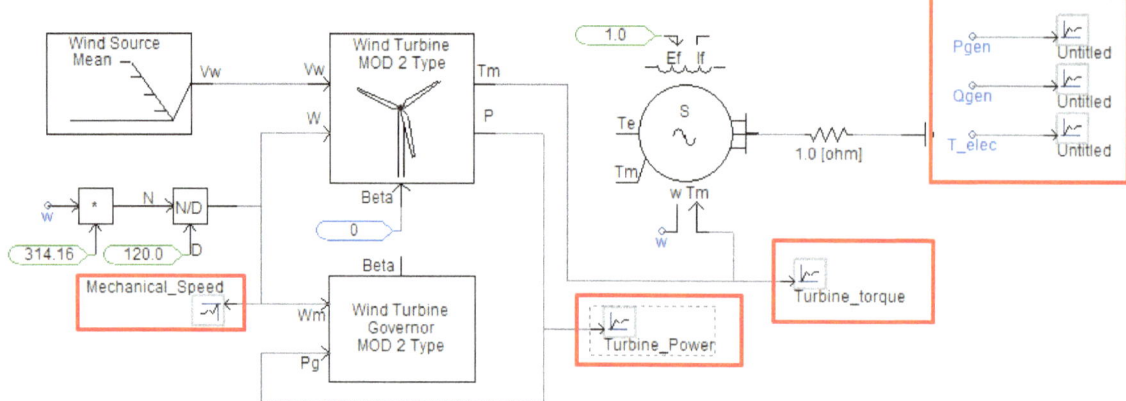

1.11.6. Simulation

For our simulation for the sake of speed and simplicity and also to the step size of the equations used in the blocks above the following parameters will be applied:

- Duration = 50 seconds
- Time step = 100 uS
- Plot step = 1000 uS
- Startup = Standard
- Plot scales = (For the Torque and Power)

For the generated active power, we will need to scale the plot with the value of 2 as we know that the generated signal from the generator is in "per unit" p.u, so we will receive 1.2 * 2 = 2.4 MW, which is the turbine power we want to achieve. The same applies for the reactive power and for the turbine power. The mechanical speed will be left as default.

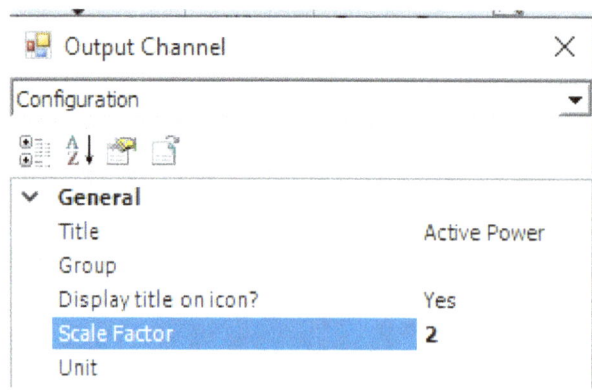

Now let's run the simulation by plotting active reactive and turbine power in a plot and the mechanical speed in the other plot.

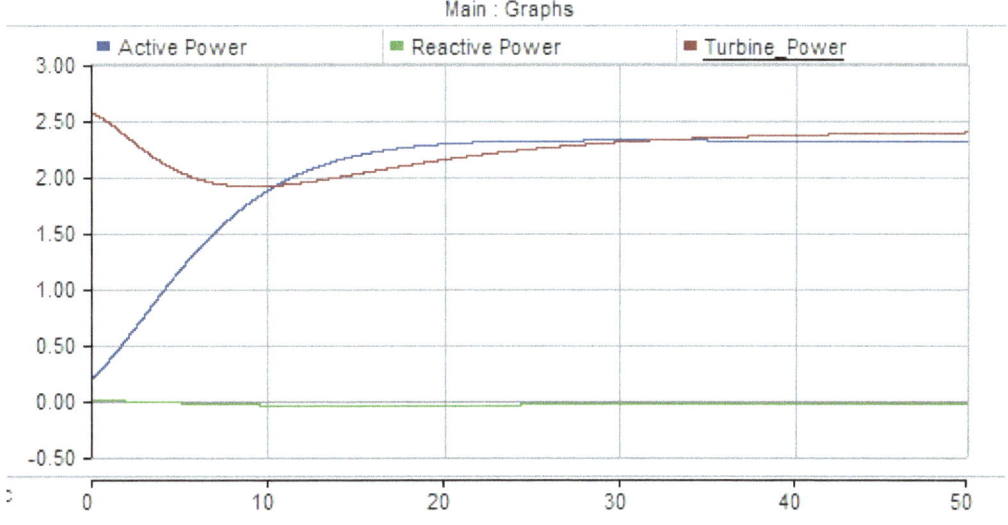

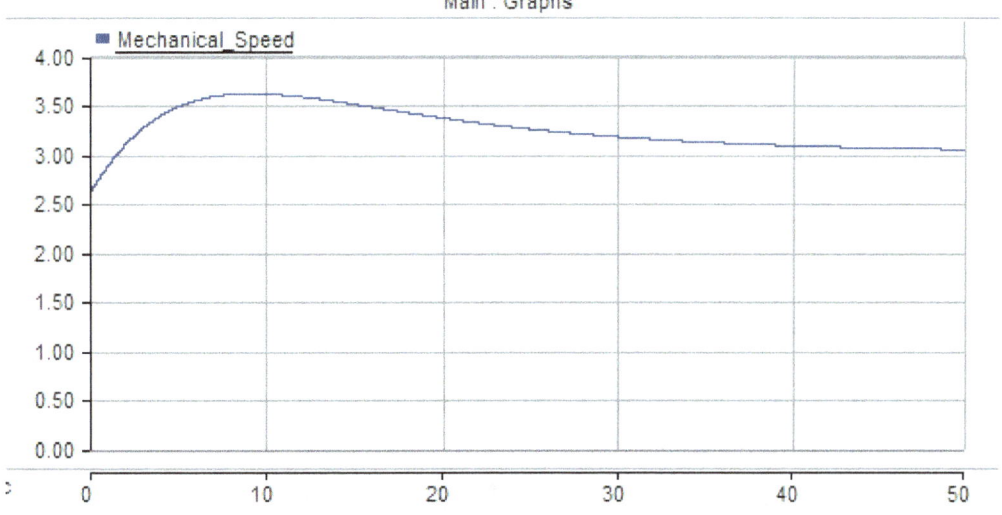

So, from the values above we have reached 2.4 MW of power.

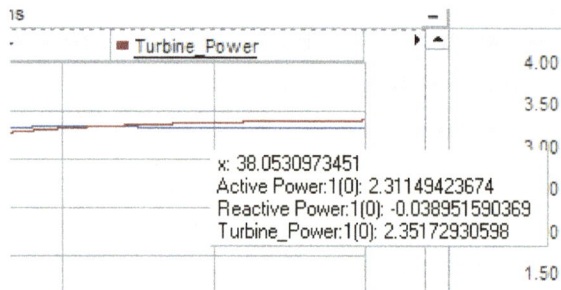

If we plot the turbine torque and electromagnetic torque as below, we obtain:

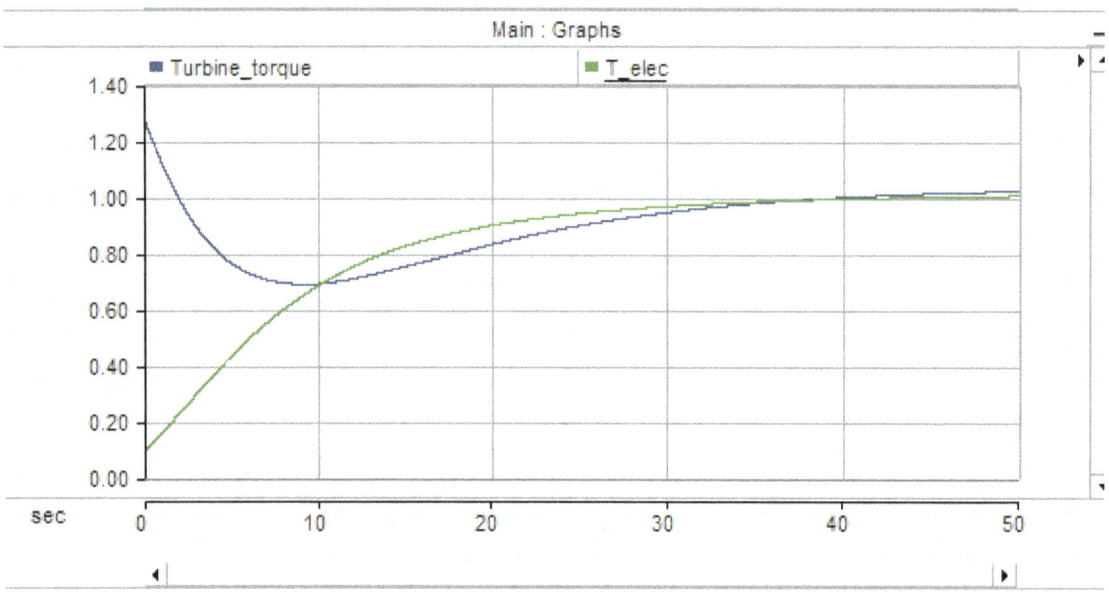

To explain the behavior above:

- From the fundamental mechanical law of torque:

$$Torque_{turbine} - Torque_{electromagnetic} = J * \frac{dw}{dt} + f * w$$

- During the starting Turbine torque > Electromagnetic torque so the speed increases
- Turbine is managed by the coefficient β which is function of wind speed and the hub speed. The wind speed is constant so when the speed is increased the β begin to decrease and the turbine power also.
- Final steady state:

$$Torque_{turbine} - Torque_{electromagnetic} = f * w$$

1.11.7.   AC -DC Power and Frequency conversion

Since the speed of the wind source is variable, an AC-DC-AC converter stage must be used to link the output of the synchronous generator—which has variable frequency and voltage—to the grid, which requires constant frequency and constant voltage. The power conversion stage will be detailed and parametrized in the portions that follow:

- Diode rectifier
- DC bus with storage capacitance voltage
- Thyristor inverter 6-pulse

The firing angle of the thyristor is not regulated by the voltage level at the grid connection point because the model only represents one wind turbine; rather, it is controlled to maintain the DC bus voltage at its rated level +/- 10%. As we shall explain later, this will include modeling HVDC control systems.

1.11.8.   Diode rectifier

A 6-pulse Graetz converter bridge (which may be an inverter or a rectifier) is included into this component, along with a Phase Locked Oscillator (PLO), firing and valve blocking controls, and measurements for firing angle (a) and extinction angle (g). Additionally, each thyristor has built-in RC snubber circuits. The 6 Pulse Bridge possesses the following external inputs and outputs:

- ComBus: Input signal to the internal Phase Locked Oscillator. This input is connected to the commutation bus though the Node Loop component as shown below:
- AO: Input alpha order (firing angle) for the converter.
- KB: Input block/deblock control signal. See Firing and Blocking Control for more details.
- AM: Measured alpha (firing angle) output [rad].
- GM: Measured gamma (extinction angle) output [rad].

This 6-pulse bridge can be used as a thyristor bridge or diode bridge but with a firing angle of 0.

This can be inserted from the figure below:

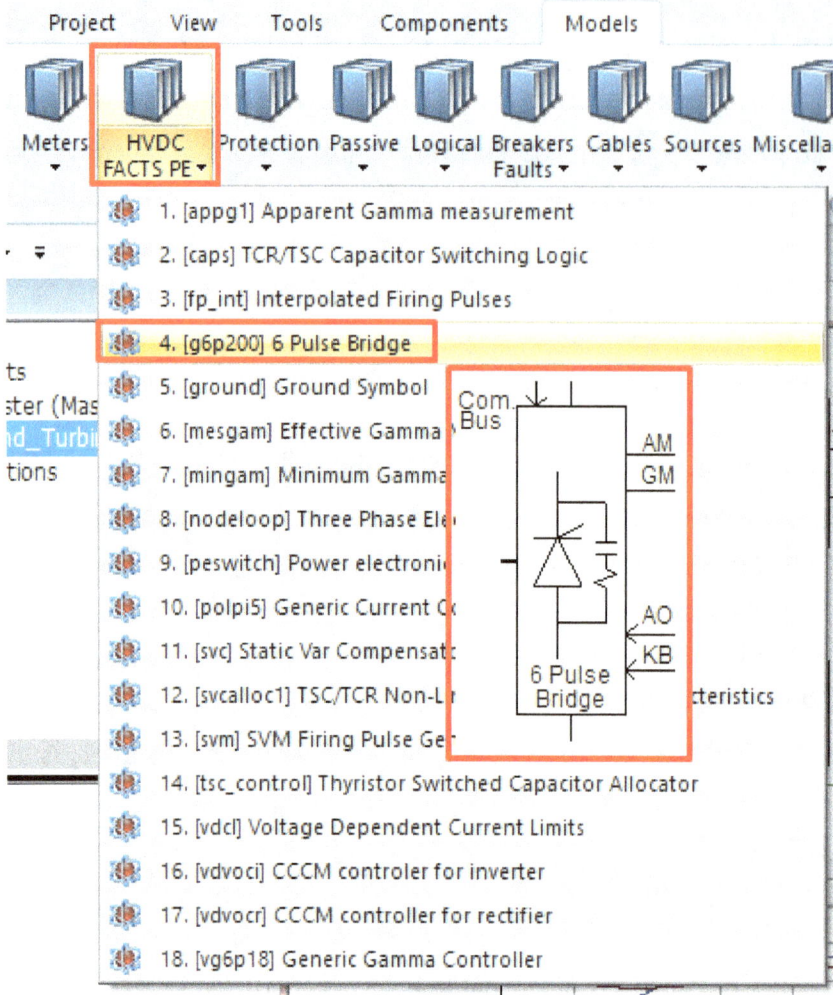

Let's connect to the wind turbine by removing the load as below:

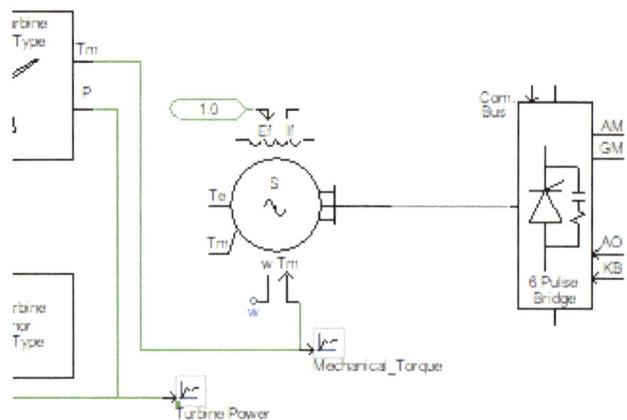

Now let's model the element as below:

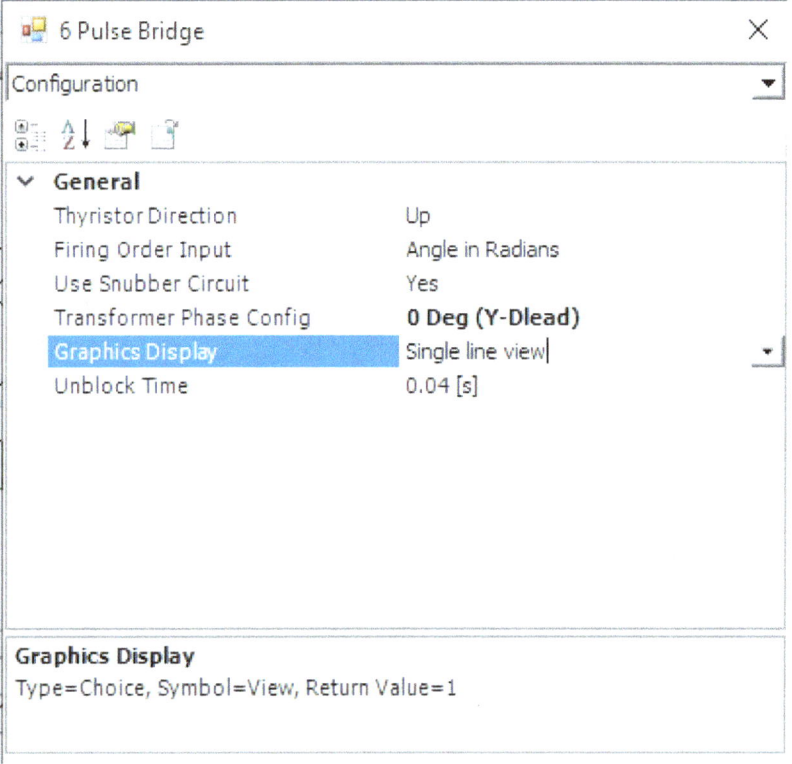

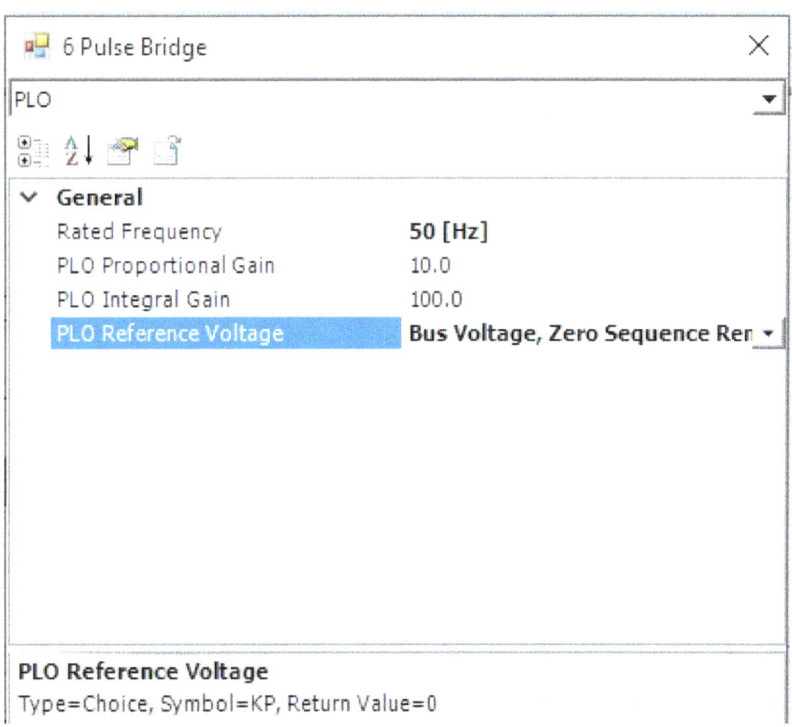

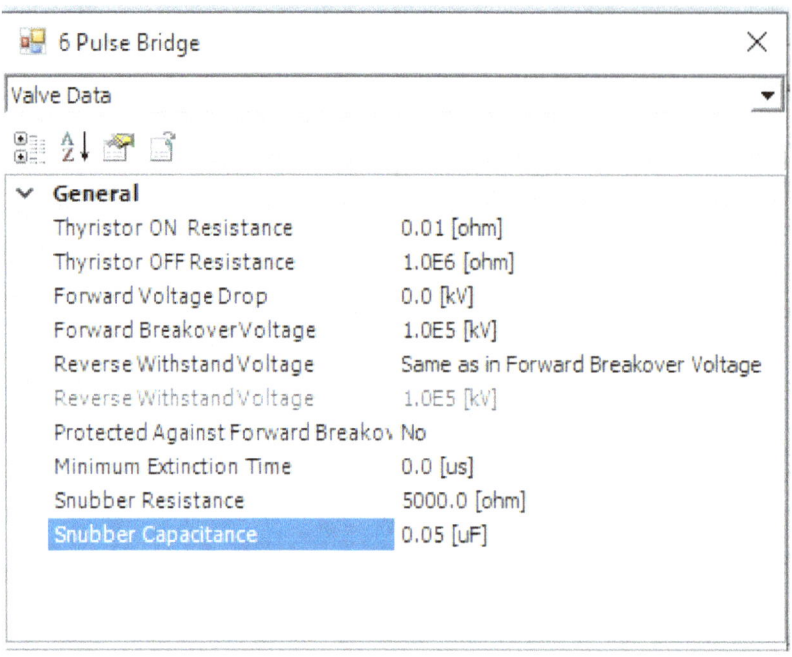

1.11.9. Overvoltage protection

The rated value of the DC voltage is calculated as below:

$$V_{Bus_{DC}} = \frac{2 * Vn * \sqrt{6}}{\pi} = \frac{2 * 400 * \sqrt{6}}{3.14} = \mathbf{624\ V}$$

The output voltage a generator is proportional to the speed that it has. We are not controlling the speed of our generator so the DC bus must be protected from overvoltage. With a margin security rate of 10% the maximum voltage is 624 +(624*10%) = **686 V**

In case of overvoltage the bus should be secured. This can be done by a comparator block that compares the signal with a condition. Under the CSMF database search for Single Input Level Comparator and insert it like in the figure below:

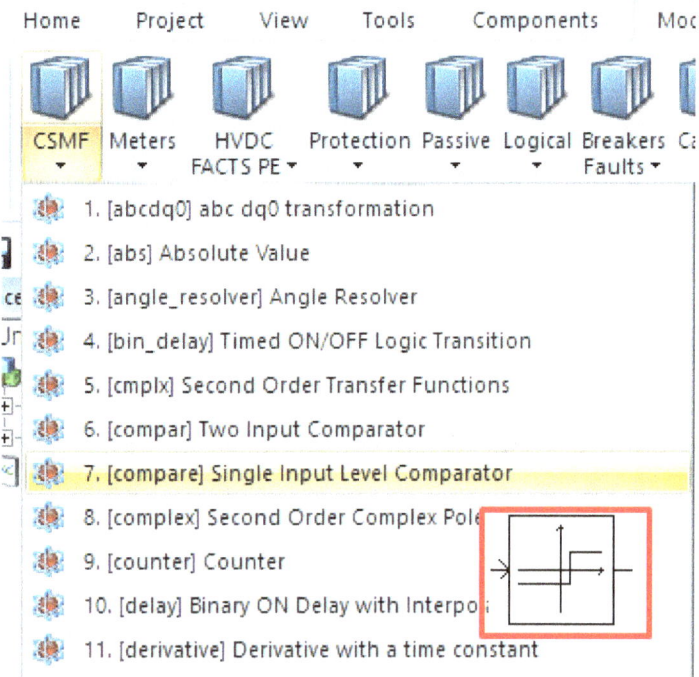

Let's configure the comparator as below:

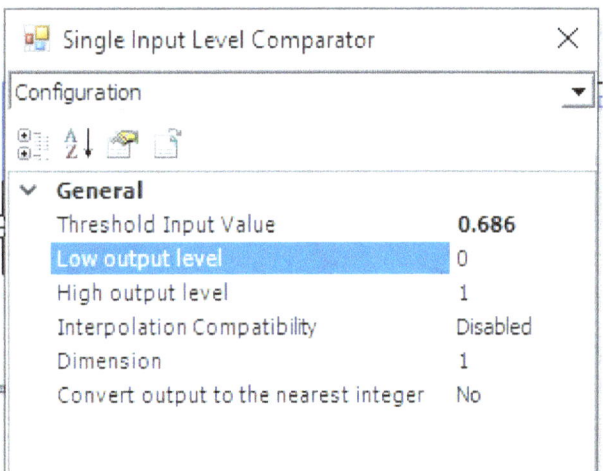

The value 0.686 corresponds to the voltage cut value of 0.686 but here unit is in kV.

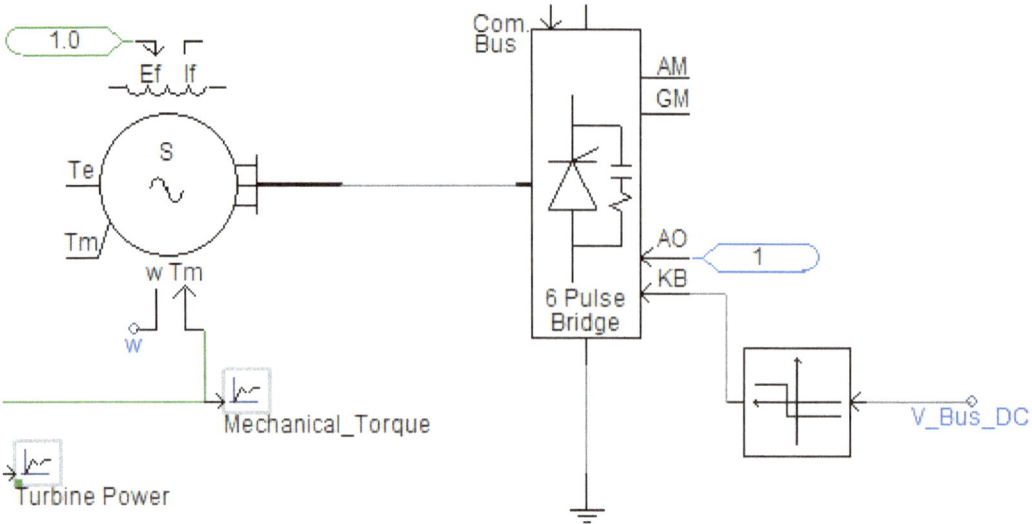

To create a communication line between the rectifier and the wind turbine insert a three-phase electrical node and connect it like in the figure below:

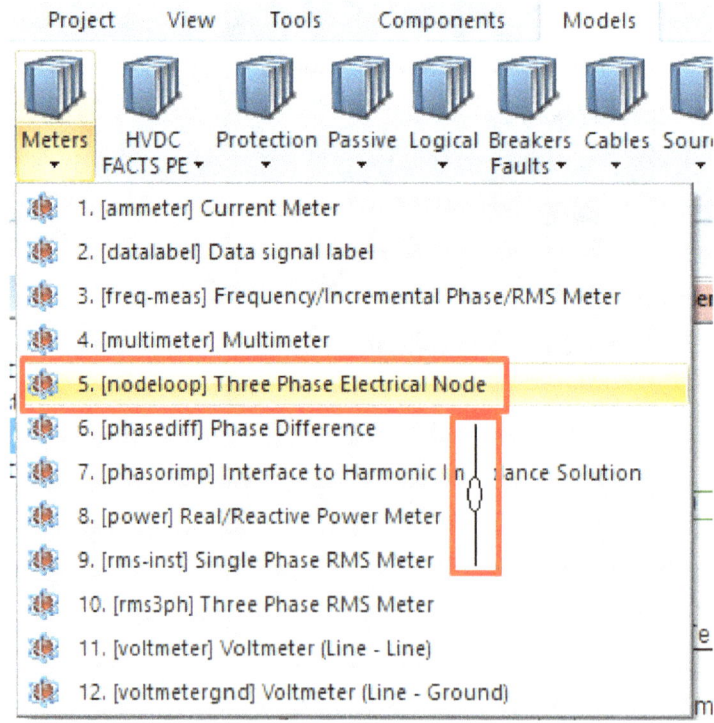

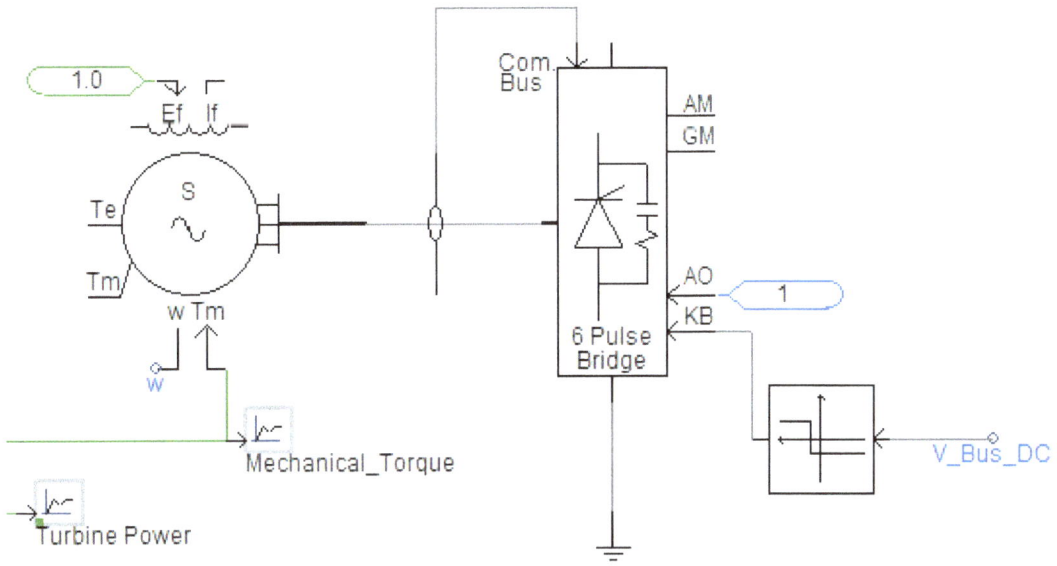

1.11.10. DC Bus

- Storage capacitor

The energy stored in the DC bus must have a tolerance of a voltage sag of 1 second.

Energy that is stored is

W= P * t = 2MW * 1s = 2 MJ

$E = 0.5 * C * V\_Bus_{DC}^2$

$C = 2 *W / V\_Bus_{DC}^2 = 2 * 2e6 / 624^2 = 10.27$ F

- Resistor

To limit the current peak from capacitor when capacitor is at low charge, we need a resistor.

V_Bus_DC = Vres + Vcap = Vres low charge

R = V_Bus_DC / I = 624/1666.6 = 0.37 Ohm

- Breaker

The system that we are using is a first order one, its load time constant is Tr = 2 * τ (τ=RC)

Tr = 2 * RC = 2*3.1*0.674 = 4.17 second

The resistor must be shunted after 4.17 seconds in order to limit Joule losses.

To perform that we need to insert a single-phase breaker (which we have learned in this course). Also, we will need to control the breaker with sequences. Insert the breaker, the capacitor and the resistor like in the figure below and let's configure them. Also make sure that you insert measuring devices for the current of the resistor and for our DC voltage.

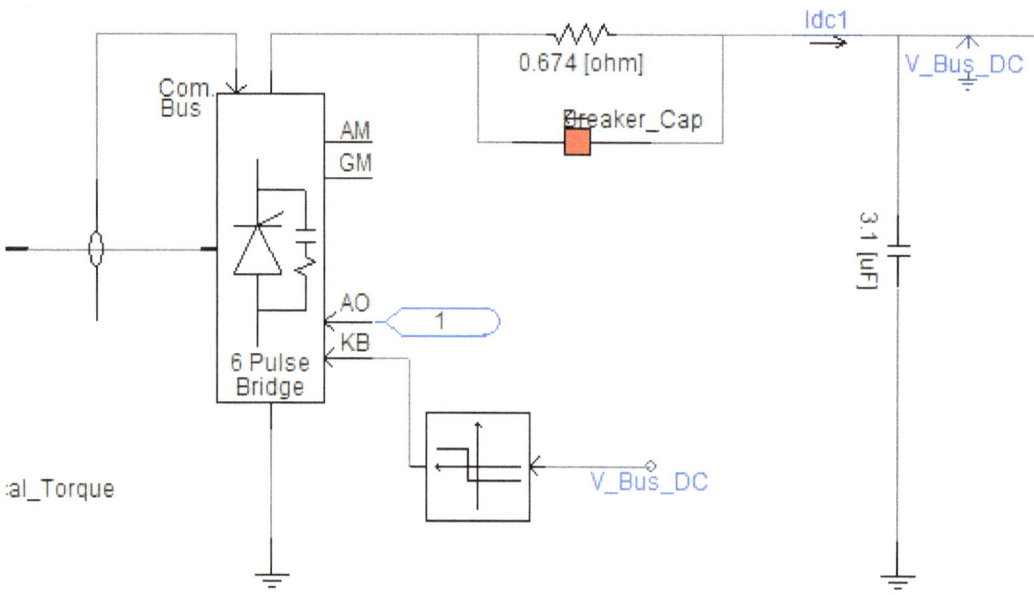

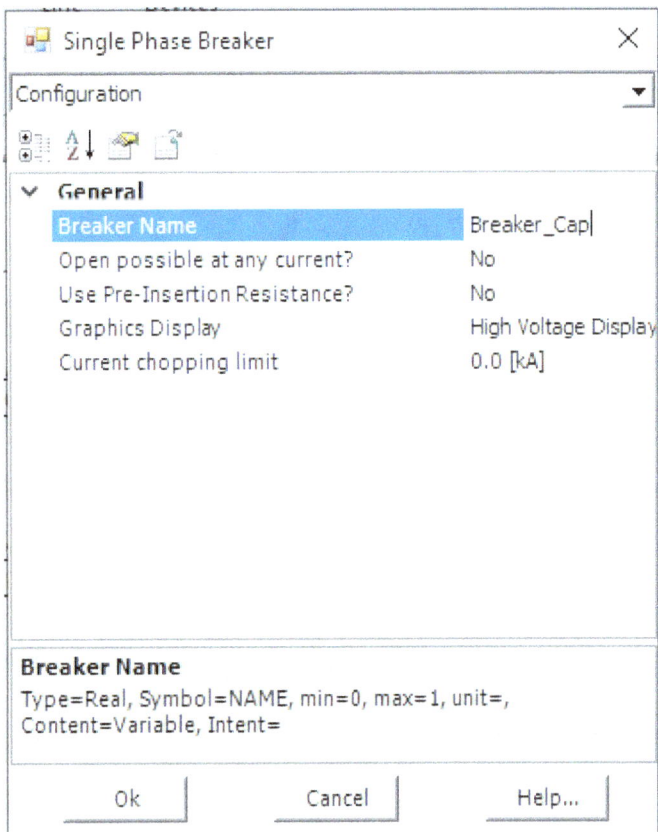

Then control the breaker with the sequences as below

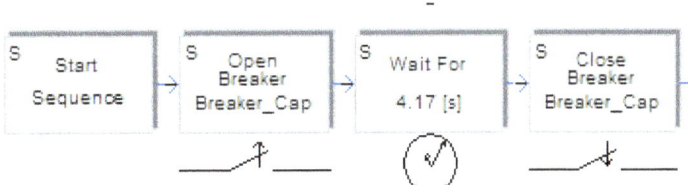

Now we will simulate the model and except the first variables that we were observing, we are going to observe the RMS values of (Root mean square) the current in the resistor and the DC bus voltage. To do this insert a Rms block like in the figure below and configure a smoothing factor of 2:

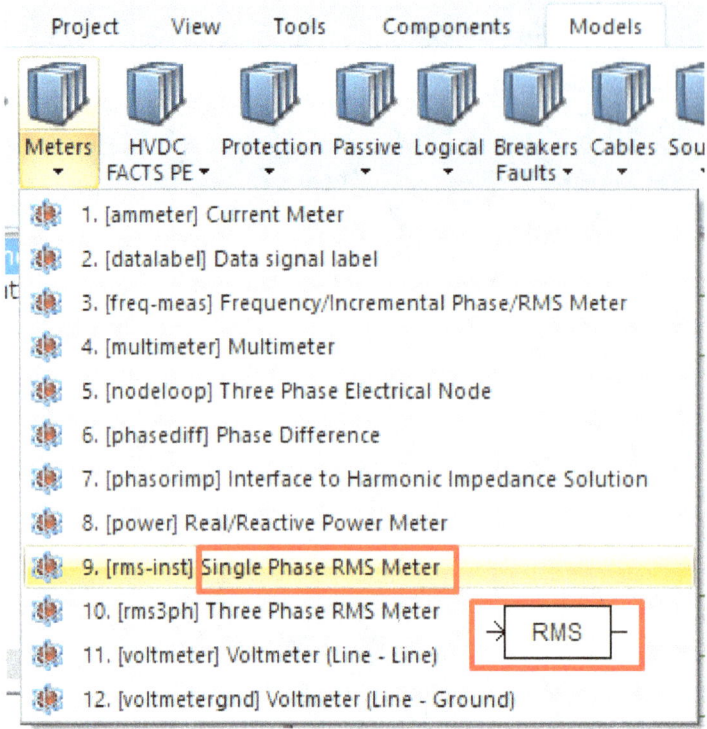

Now let's run the simulation with the same configuration and with a time duration of 60 seconds.

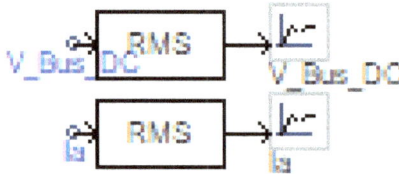

Now let's try to simulate the system for 60 seconds and with a RMS smoothing constant of 2 seconds.

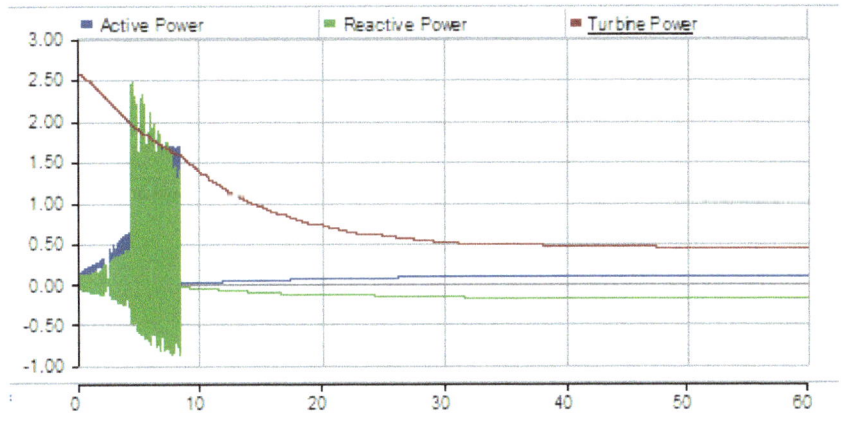

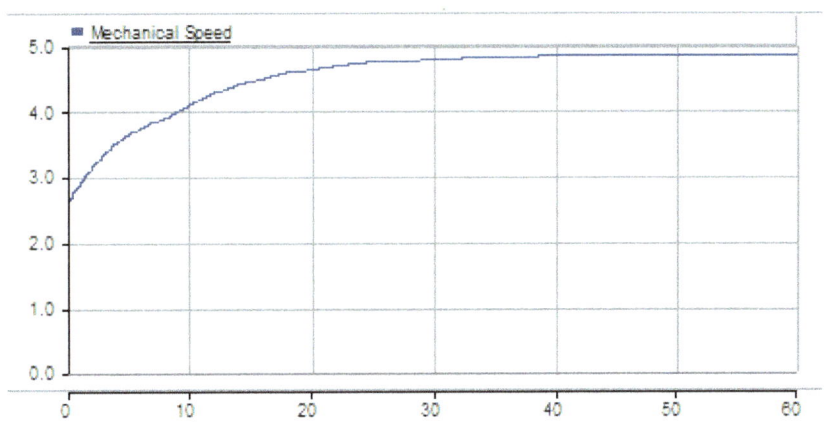

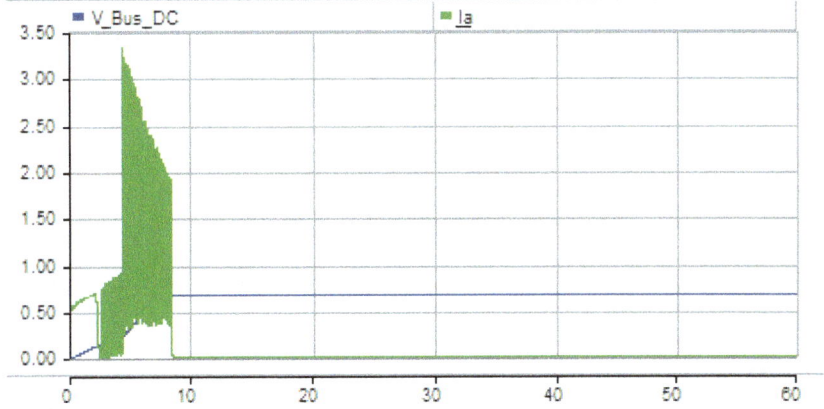

The speed rises because the motor torque is always greater than the resistive torque. The turbine power and torque decrease as the speed rises because the beta factor falls. The overvoltage regulation restricts the bus to 686 volts at the moment the breaker is closed, shunting the resistance, charging the capacitor, and bringing the current Idc to zero. In this manner, the wind turbine's output power is reduced to zero.

1.11.11. Inverter – 6 pulses

The inverter that we modeled is a current one (monodirectional). To model a current source as its input, we need to include an inductor. Like the capacitor, also in the inductor the energy must be enough to store energy for 1 second.

W = 2000000 J

The energy stored is W = $0.5 * L * Idc^2$

We know that Idc = Pdc / Vdc = 2000000/624 = 3205 A

Then L = $2*E/ Idc^2$ = $2*2000000/3205^2$ = 0.39 H

Insert a inductor of 0.39 H in the diagram as below:

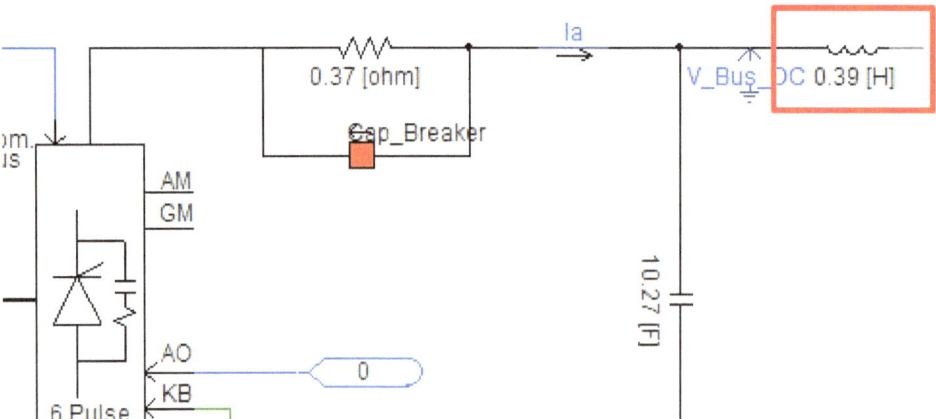

1.11.12. Inverter

To obtain current thyristor inverter insert the 6-pulse thyristor converter and configure like in the figure below:

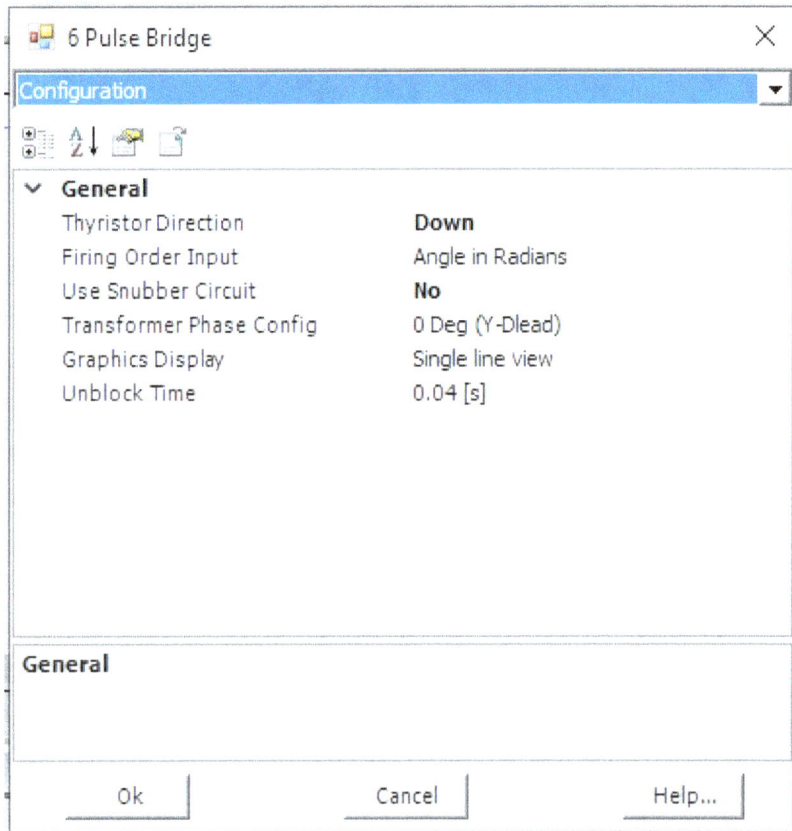

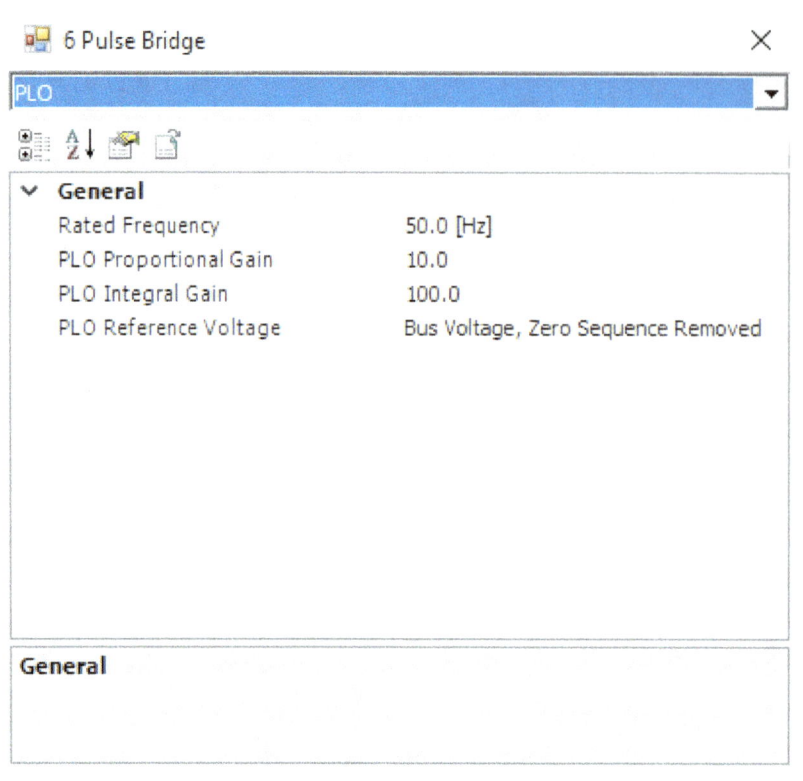

6 Pulse Bridge   ✕

PLO ▼

✓ General

Rated Frequency	50.0 [Hz]
PLO Proportional Gain	10.0
PLO Integral Gain	100.0
PLO Reference Voltage	Bus Voltage, Zero Sequence Removed

General

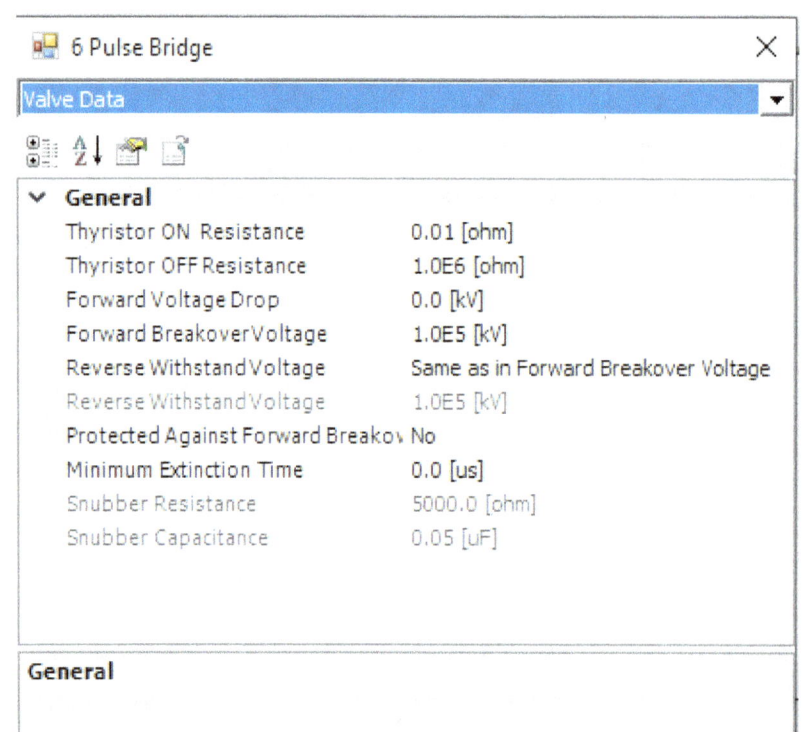

6 Pulse Bridge   ✕

Valve Data ▼

✓ General

Thyristor ON Resistance	0.01 [ohm]
Thyristor OFF Resistance	1.0E6 [ohm]
Forward Voltage Drop	0.0 [kV]
Forward Breakover Voltage	1.0E5 [kV]
Reverse Withstand Voltage	Same as in Forward Breakover Voltage
Reverse Withstand Voltage	1.0E5 [kV]
Protected Against Forward Breakov	No
Minimum Extinction Time	0.0 [us]
Snubber Resistance	5000.0 [ohm]
Snubber Capacitance	0.05 [uF]

General

Insert the inverter (rotate and flip) to look like in the figure below:

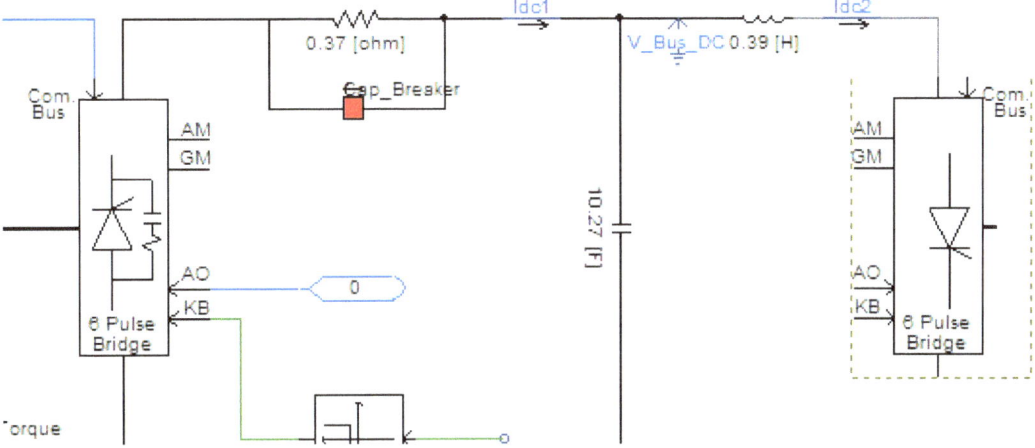

The inverter must provide voltage control and voltage collapse protection for the DC bus.

1.11.13. Voltage collapse limitation

For the low voltage bus limitation, we will need a 10 percent less limit

V_Bus_DC = 0.9*624 = 562 V

Then insert this value in a comparator and configure and connect it like below:

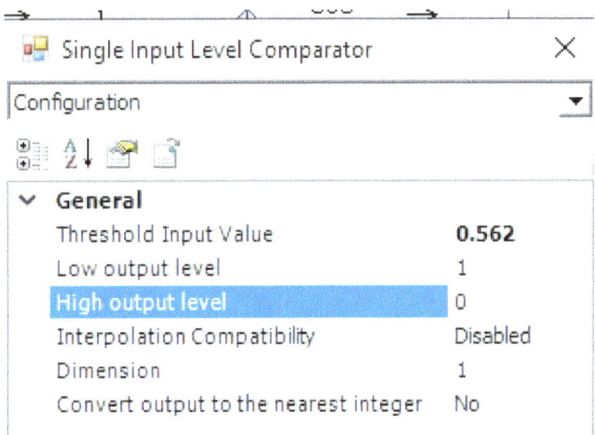

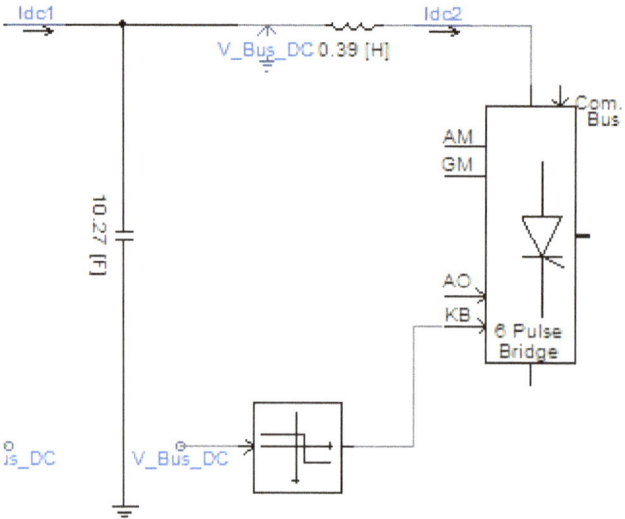

1.11.14. Voltage Regulation

We are modeling a single wind turbine and before connecting this wind turbine to the grid we must emphasize the fact that the control of the voltage will be done at the DC bus rather at the connection point. This is due to the fact that we have just one wind turbine and the impact at the grid is weak. So, the fluctuation of the DC bus will be inside 0.95 pu and 1.05 pu.

$624 * 0.85 < V\_DC\_Bus < 624 * 1.05$

$593 < V\_DC\_Bus < 655$

To control the voltage, we will need two components from PSCAD library

- Voltage dependable current limits

This component generates a voltage-dependent current restriction that is either started by a timer that starts when the measured DC volts drop below a certain level or by a lag function. The input parameter Delay or Lag Function determines this.

The majority of DC transmission systems must reduce the impact of heavy currents delivered to or pulled from AC systems with disproportionately little actual power. The converters' AC voltage will sympathetically fall with low short circuit ratio. In DC networks, voltage-dependent current limitations are frequently utilized to prevent the continued operation of such circumstances. Recognizing the collapse of the DC voltage, the current order to the current controller of each converter is decreased to within acceptable bounds, say 0.2 to 0.5 per unit.

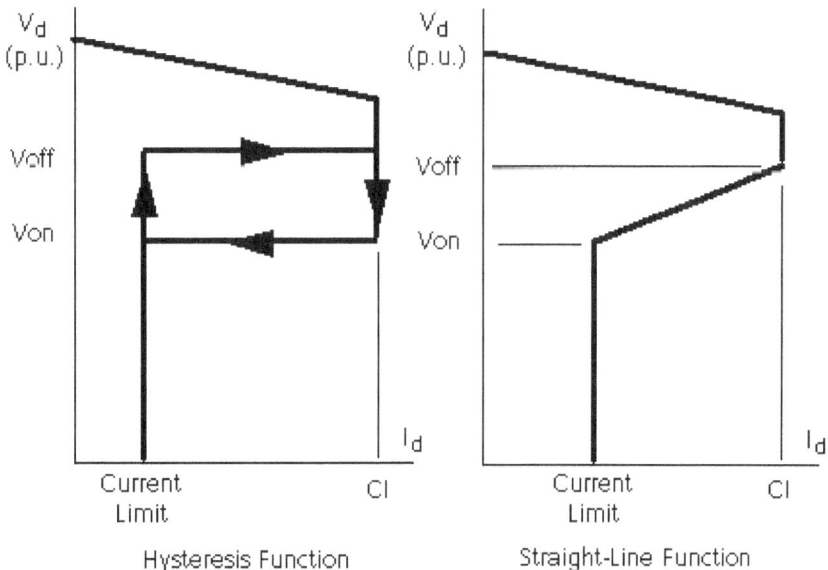

Hysteresis Function   Straight-Line Function

The Voltage Dependent Current Limits component possesses the following external inputs and outputs:

- **VD:** DC side measured volts (negative for negative pole) [kV]
- **CI**: Current order (always positive) [pu]
- **CO**: Current order output (always positive) [pu]

We will define two voltage levels and we want the DC voltage to operate within this range. Those values are Applying Limit (Von) and Removing Limit (Voff). The user also enters a minimum value for the current, called "Current Limit" so:

- If VD > Removing Limit then Current Order CO = Current Input CI
- If VD < Applying Limit then Current Order CO = Current Limit

Let's insert the block and configure as below:

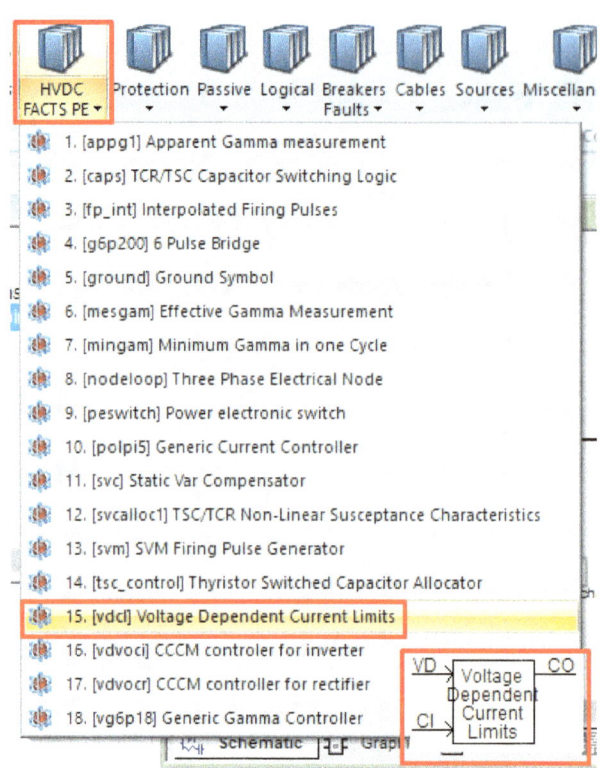

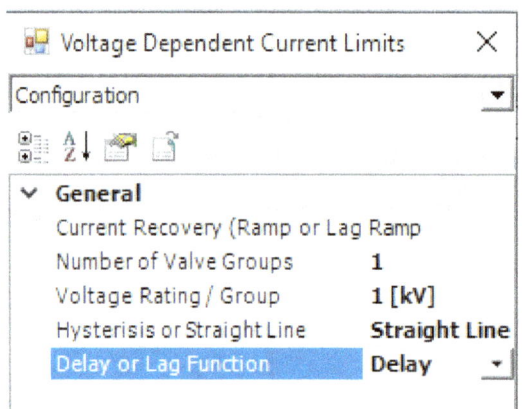

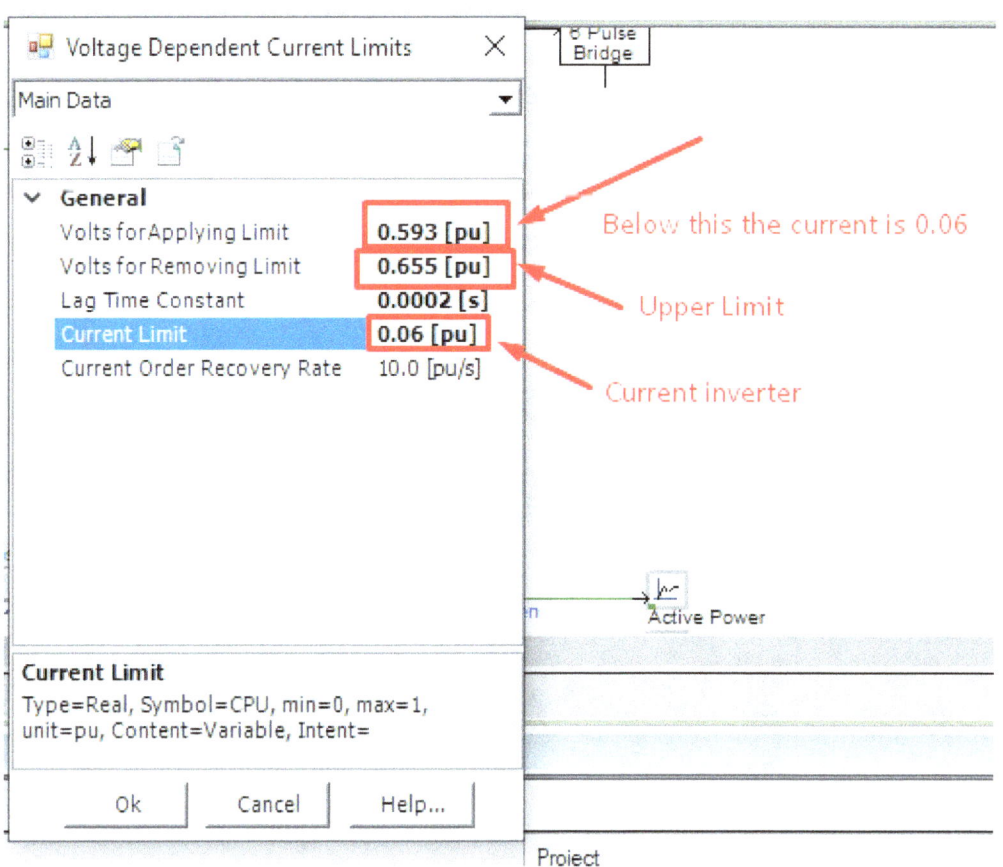

Voltage Dependent Current Limits

Main Data ▾

∨ **General**
Volts forApplying Limit   **0.593 [pu]**   Below this the current is 0.06
Volts for Removing Limit   **0.655 [pu]**   Upper Limit
Lag Time Constant   **0.0002 [s]**
Current Limit   **0.06 [pu]**   Current inverter
Current Order Recovery Rate   10.0 [pu/s]

Current Limit
Type=Real, Symbol=CPU, min=0, max=1,
unit=pu, Content=Variable, Intent=

Ok   Cancel   Help...

6 Pulse
Bridge

Active Power

Project

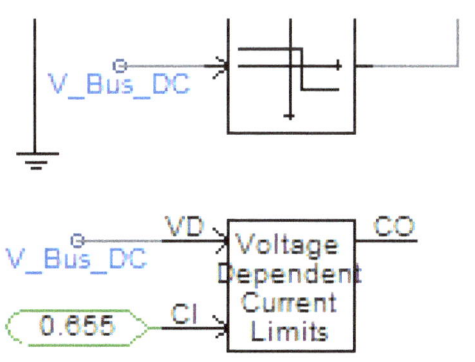

- Generic current controller

In actual HVDC systems, system protection requirements like Voltage Dependent Current Limits may be used to process the two inputs to this component, current order and intended gamma. However, as valves have a limited capacity for over-current, current management continues to be a crucial component of a DC link's performance. To make sure converter current stays at safe operating levels, current order limitations are established. Each series valve group in a pole typically receives a desired firing angle alpha from the current controller. The current controller is hence sometimes referred to as the "pole controller." To achieve steady state and transient power regulation and system protection, additional controls, protections, and restrictions are used to govern the current order into the current controller.

This allows us to produce alpha from a PI controller acting from the error between current order (CO) coming from Voltage Dependent Current Limits

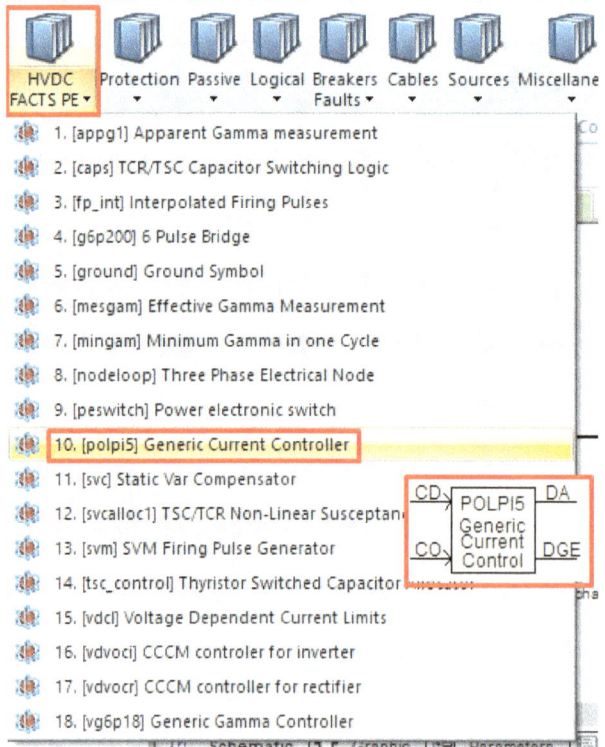

Insert this block in the model and configure as below:

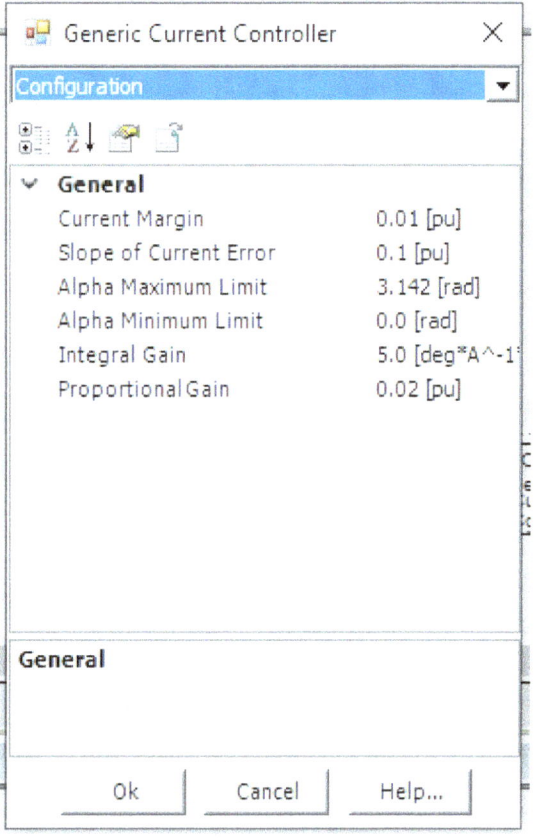

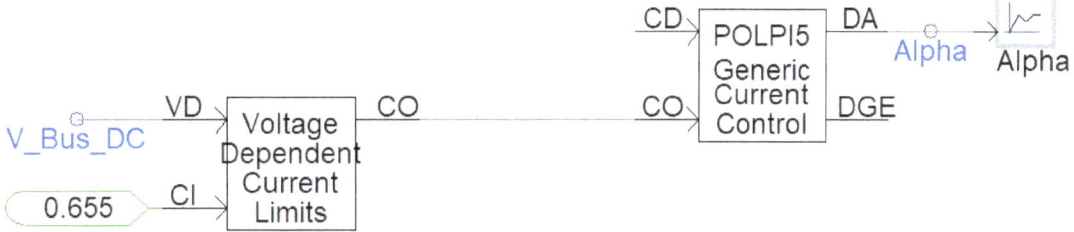

Connect elements and make sure that ground is removed and the loop is closed and nothing is connected to the ground.

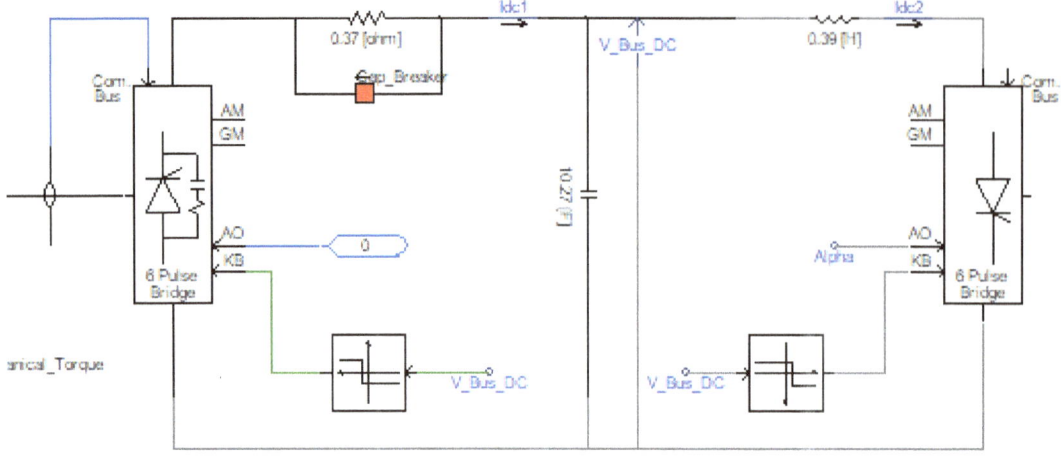

1.11.15. Connection to the grid

The last phase of the electrical grid, known as the distribution grid, is responsible for distributing power to businesses, residences, and other end users. In addition to providing electricity to every user on the grid, distribution also decreases power to levels that are safe for customers to consume. Step-down transformers or step-up transformers are required usually to connect the generation units to the required level of voltage. Before connecting to the grid, we would need the transformers and capacitors.

- Capacitors

To smooth the voltage at the connection point and compensate the reactive power. We will install a capacitor of 2mF just for the sake of simulation.

- Transformer

To adapt the connection voltage level, we would need a transformer to connect. The transformer will be connected after the inverter and we will assume the losses 0 for the inverter.

Current is ➔ Ieff = $Idc * \dfrac{\sqrt{6}}{\pi}$

Considering no losses at the inverter we have

$$P = 3 * V * Ieff * cos\emptyset$$

$$Vdv * Idc = 3 * V * Idc * cos\emptyset * \sqrt{6}/\pi$$

$$V = \pi * \frac{Vdv}{\left(3 * \sqrt{6} * cos\emptyset\right)}$$

In the thyristor, the current should be ahead of voltage so we will go for a safe margin of

$$\emptyset = \pi/4$$

From the above equations:

$$V = 3.14 * \frac{655}{3*\sqrt{6}*cos\left(\frac{\pi}{4}\right)} = 400$$

Now insert a three phase (2 windings) transformer and also the capacitor and connect like below:

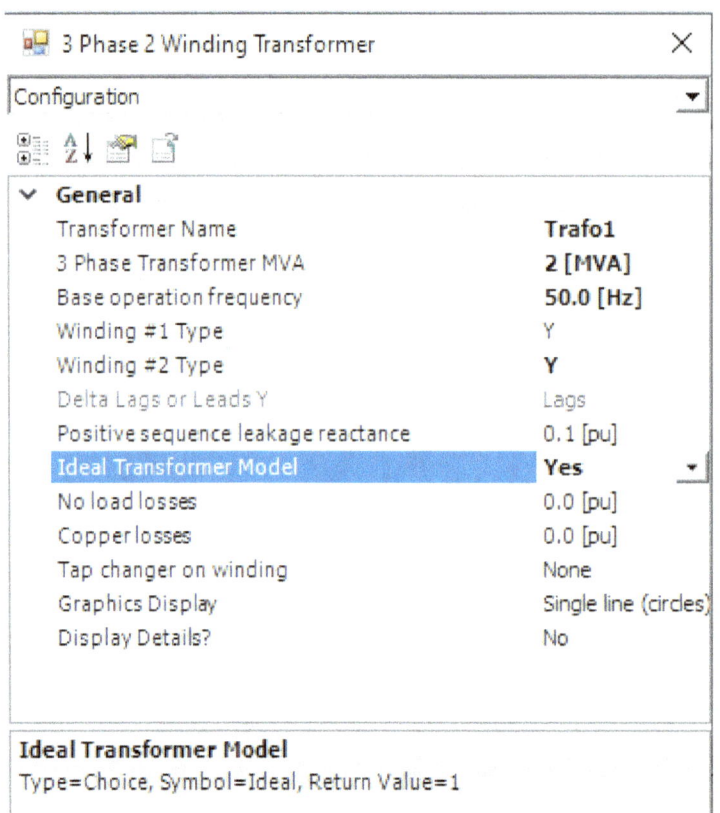

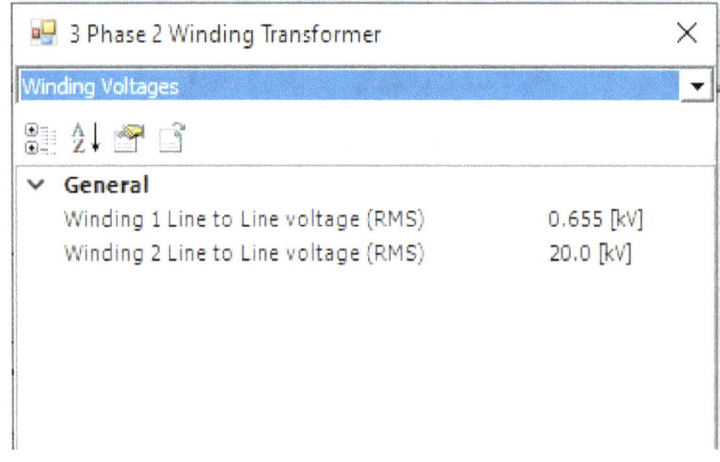

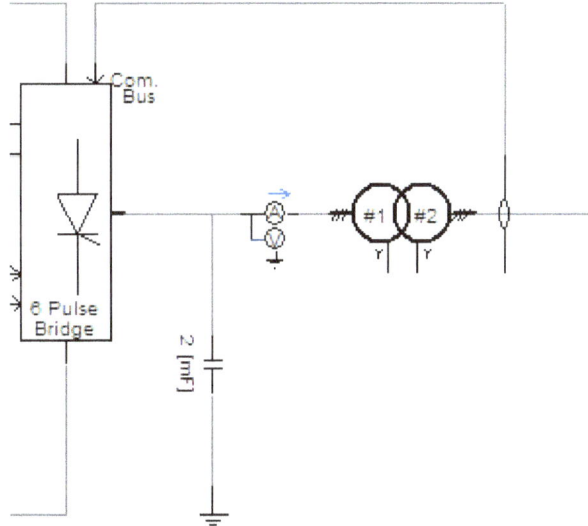

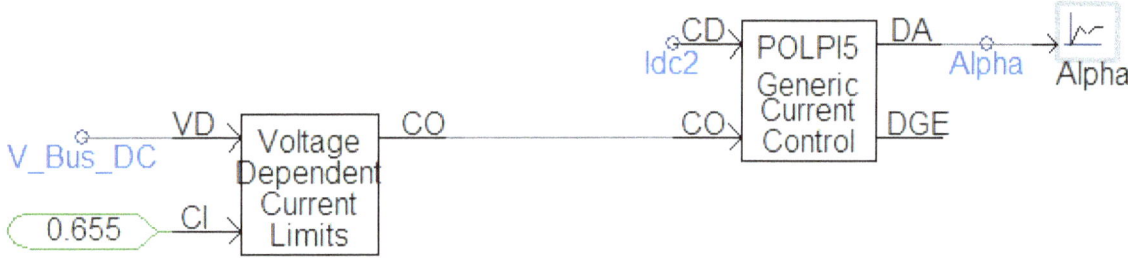

1.11.16. Distribution grid

Now we will create the distribution grid. The grid as we know on this course can be modelled simply as a three-phase source (ideal or not ideal). For our model we will create a more complex (real) distribution grid.

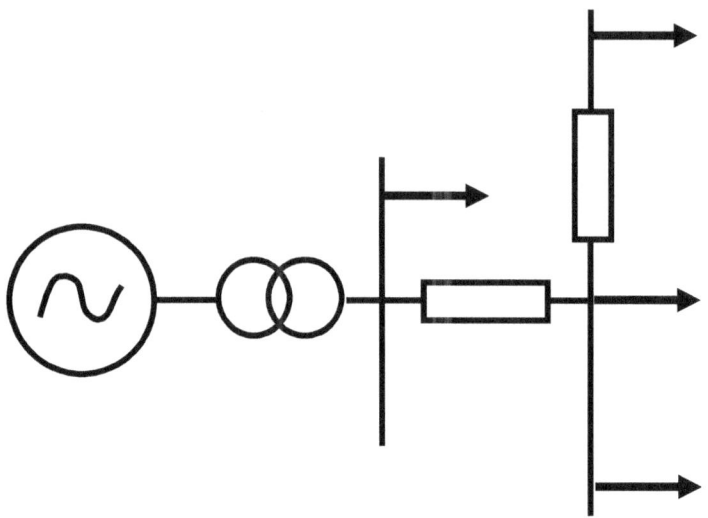

Let's try to model this in PSCAD as below:

- 3 Phase Voltage Source

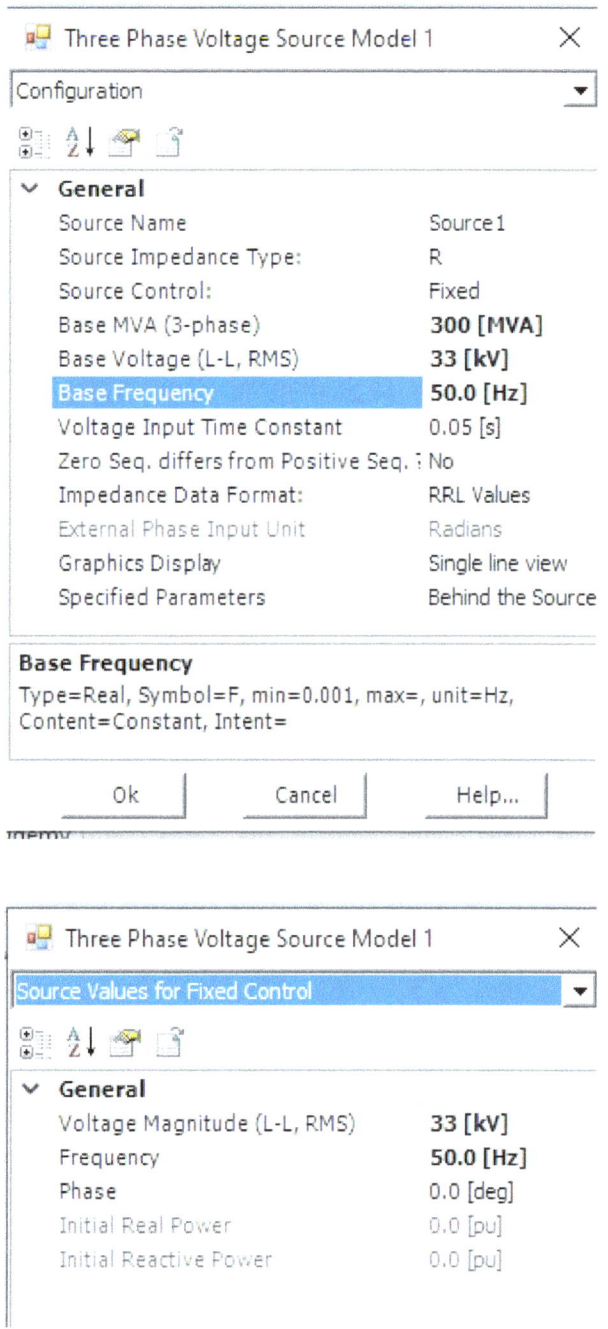

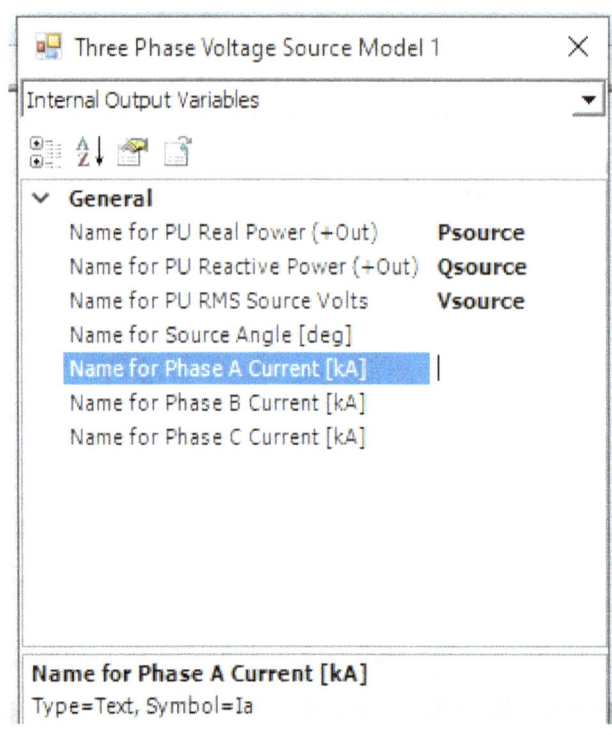

Three Phase Voltage Source Model 1   ✕

Internal Output Variables ▾

⌄ **General**
Name for PU Real Power (+Out)	**Psource**
Name for PU Reactive Power (+Out)	**Qsource**
Name for PU RMS Source Volts	**Vsource**
Name for Source Angle [deg]	
Name for Phase A Current [kA]	
Name for Phase B Current [kA]	
Name for Phase C Current [kA]	

Name for Phase A Current [kA]
Type=Text, Symbol=Ia

- Transformer

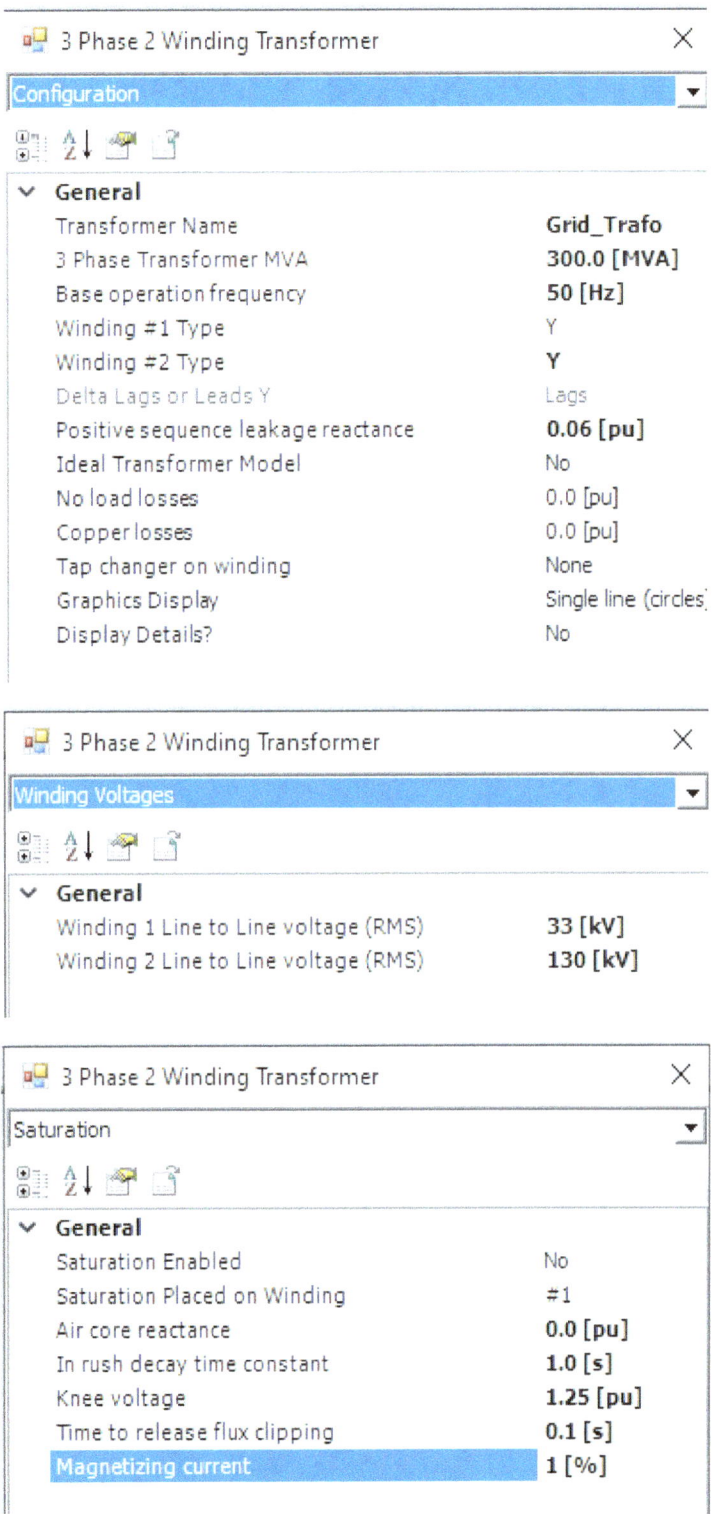

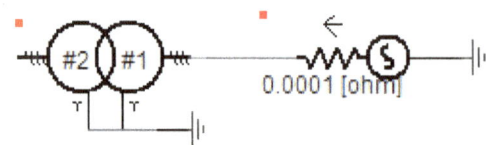

- Nodes and Loads

Insert Load at node 1:

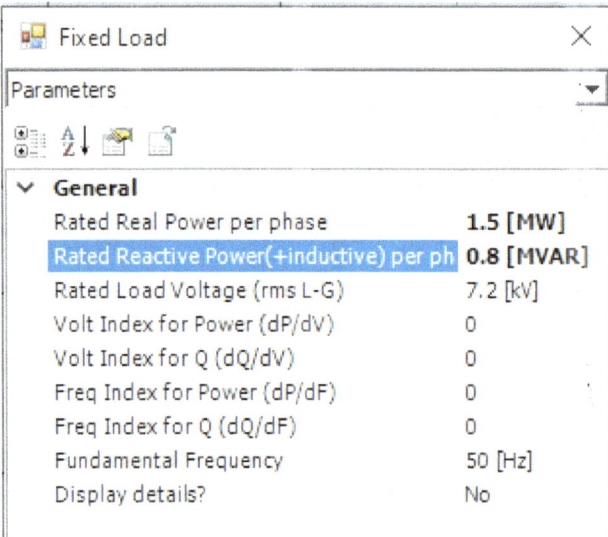

Then it should look like below

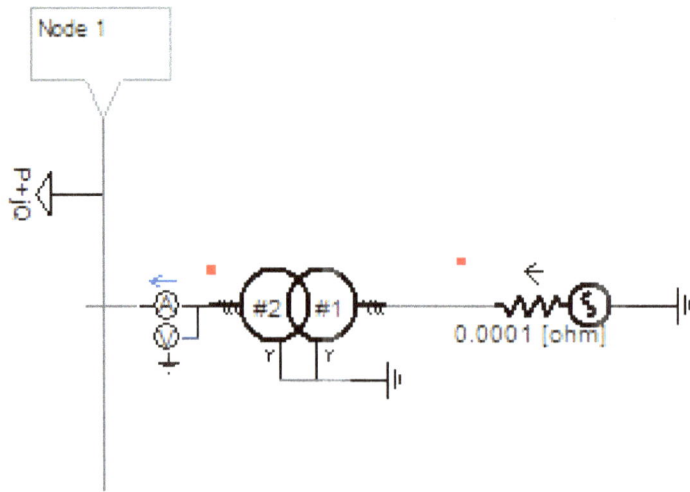

Then the loads for node 2 and 3:

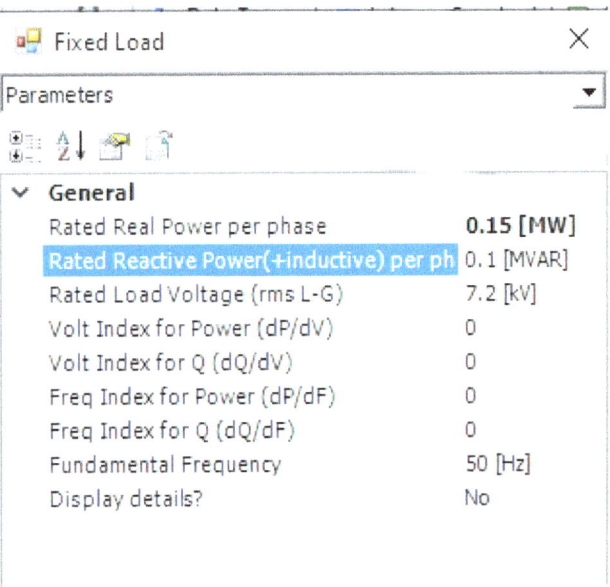

And the final view should look like:

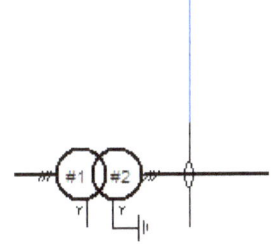

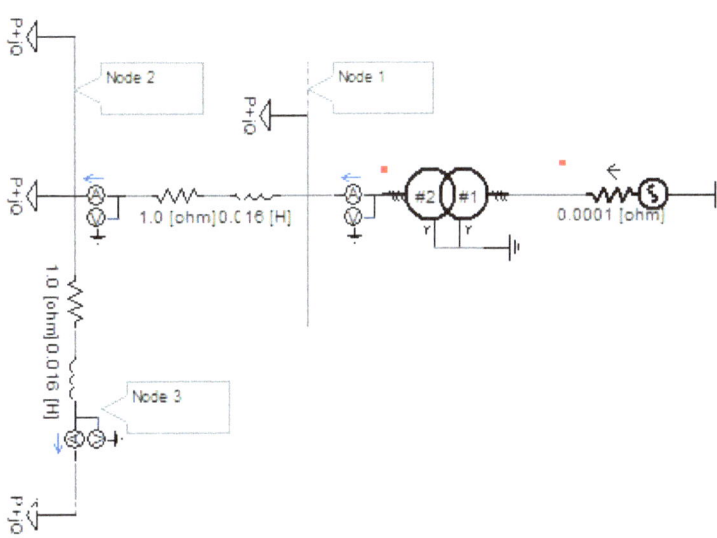

1.11.17. Measurements

Let's measure some relevant signals

- From all the measurements in the grid we extract the currents with a RMS 0.2 smoothing factor:

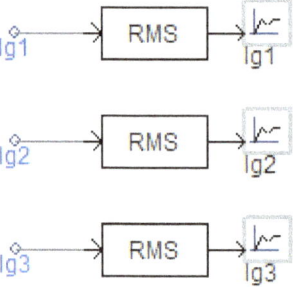

1.11.18. Simulation

Now let's simulate the whole park by connecting the wind turbine in one of our nodes. We will simulate by connecting at the node 2.

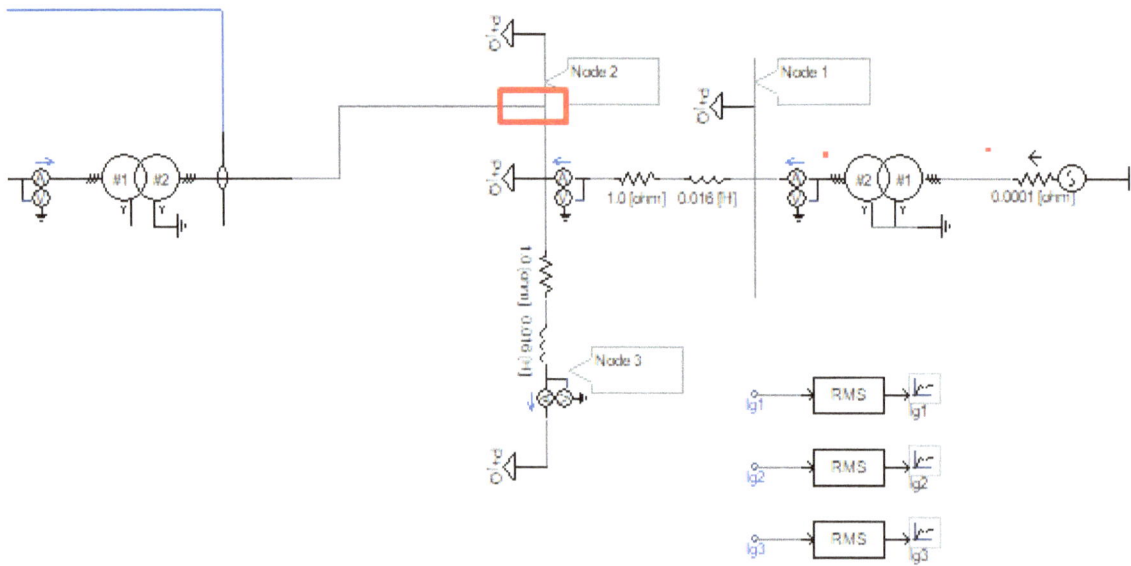

POWER TO HUMAN